Edexcel GCSE Mathematics A Linear

Foundation REVISION WORKBOOK

Series Director: Keith Pledger
Series Editor: Graham Cumming

Authors: Julie Bolter, Gwenllian Burns, Jean Linsky

The Edexcel Revision Series

These revision books work in combination with Edexcel's main GCSE Mathematics 2010 series. The Revision Guides are designed for independent or classroom study. The Revision Workbooks use a write-in format to provide realistic exam practice.

Specification A Linear **Specification B Modular**

Higher

Foundation

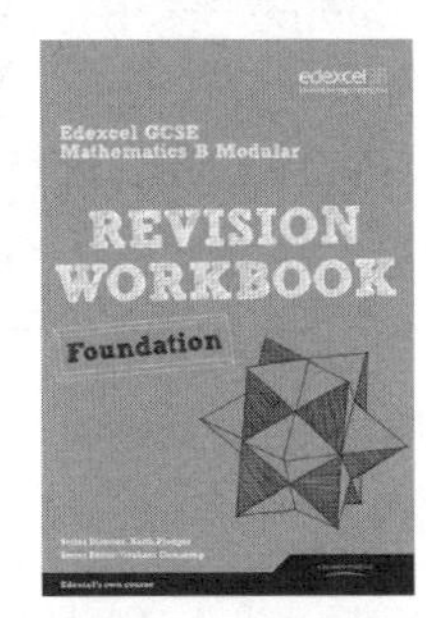

A PEARSON COMPANY

Contents

A small bit of small print

A grade allocated to a question represents the highest grade covered by that question. Sub-parts of the question may cover lower grade material.

The grade range of a topic represents the usual grade range that the topic is assessed at. The topic may form part of a higher grade question if tested within the context of another topic.

Questions in this book are targeted at the grades indicated.

Place value

G 1 (a) Write the number 7082 in words.

You can use a place value diagram to help you read and write numbers.

Thousands	Hundreds	Tens	Units
7	0	8	2

.... **(1 mark)**

(b) Write 4670 to the nearest hundred.

.... **(1 mark)**

(c) Write down the value of the **5** in the number 6759

.... **(1 mark)**

(d) Write the number **three thousand, one hundred and seventeen** in figures.

.... **(1 mark)**

G 2 43 290 people watched a football match.

(a) Write 43 290 to the nearest thousand.

.... **(1 mark)**

(b) Write down the value of the **3** in the number 43 290

.... **(1 mark)**

G 3 Write these numbers in order of size. Start with the smallest number.

(a) 784 470 78 84 478

Guided

~~784~~, 470, ~~78~~, 84, 478

78 .... 784 **(1 mark)**

(b) 5431 3451 4531 1435 1453

.... **(1 mark)**

G 4 (a) Write **four million** in figures.

.... **(1 mark)**

(b) (i) Write 350 000 in words.

.... **(1 mark)**

(ii) Write down the value of the **8** in the number 7834

.... **(1 mark)**

(c) (i) Write 38 920 correct to the nearest thousand.

.... **(1 mark)**

(ii) Write 38 920 correct to the nearest hundred.

.... **(1 mark)**

G 5 Write these numbers in order of size. Start with the smallest number.

78 900 79 000 781 900 780 000 98 000

.... **(1 mark)**

Rounding numbers

G **1** (a) Round 4878 to the nearest thousand.

.... **(1 mark)**

(b) Round 23 419 to the nearest hundred.

.... **(1 mark)**

(c) Round 6729 to the nearest ten.

.... **(1 mark)**

E **2** (a) Write the number 46 784 correct to 1 significant figure.

Guided

4 | 6784 = ... correct to 1 significant figure

Round the first figure. Remember to replace the 6784 with 0000.

(1 mark)

(b) Write the number 0.056 921 correct to 1 significant figure.

Guided

0.05 | 6921 = ... correct to 1 significant figure

The first significant figure is the 5.

(1 mark)

D **3** (a) Write the number 3.4982 correct to 1 significant figure.

.... **(1 mark)**

(b) Write the number 70 319 correct to 2 significant figures.

.... **(1 mark)**

(c) Write the number 0.089 53 correct to 2 significant figures.

.... **(1 mark)**

D **4** Round 7.0263 correct to

(a) 1 significant figure

.... **(1 mark)**

(b) 2 significant figures

.... **(1 mark)**

(c) 3 significant figures

.... **(1 mark)**

D **5** Round 0.009 275 correct to

(a) 1 significant figure

.... **(1 mark)**

(b) 2 significant figures

.... **(1 mark)**

(c) 3 significant figures

.... **(1 mark)**

Adding and subtracting

G **1** Work out

(a) 673 + 89 + 34

Guided

```
    6 7 3
      8 9
+     3 4
  -------
  ......6
      1
```

(1 mark)

(b) 561 − 87

```
  5 ⁵6̸ ¹1
−     8 7
  -------
  ......4
```

(1 mark)

G **2** Work out

(a) 6540 + 385

(1 mark)

(b) 4308 − 842

(1 mark)

G **3** Work out

(a) 385 + 64 + 1023

(1 mark)

(b) 438 – 295

(1 mark)

G **4** Jodi has £205 in a savings account.
She spends £119 on an MP3 player.
How much money does she have left in her account?

£.................... **(1 mark)**

G **5** There are 45 people on a coach.
34 people get off the coach.
23 people get on the coach.
How many people are now on the coach?

45 − 34 = ...

...... + 23 = ...

(2 marks)

F **6** 27 people were on a bus.
At the first stop, 18 people got off and 12 people got on.
At the second stop, 19 people got off and 25 people got on.
How many people are now on the bus?

.... **(3 marks)**

Multiplying and dividing

G **1** Work out

(a) 46×30

Guided

```
      4 6
  ×   3 0
  -------
  .......
```

(1 mark)

(b) $513 \div 3$

```
    ......
  3)5 1 3
```

(2 marks)

G **2** Fatima bought 48 teddy bears at £9 each.
Work out the total amount she paid.

£.................... **(1 mark)**

F **3** A box of chocolate bars contains 26 bars.
Work out the total number of bars in 34 boxes.

.... **(3 marks)**

E **4** Work out

Exam questions similar to this have proved especially tricky – be prepared! ResultsPlus

(a) 564×35

Guided

EXAM ALERT

```
      5 6 4
  ×     3 5
  ---------
  .........   Work out 564 × 5
  .........   Work out 564 × 30
  ---------
  .........
```

(3 marks)

(b) $425 \div 17$

.... **(3 marks)**

E **5** A restaurant shares £345 in tips equally between 15 members of staff.
How much does each person get?

£.................... **(3 marks)**

Decimals and place value

G **1** (a) Write down the value of the **6** in the number 8.62

.... **(1 mark)**

(b) Write down the value of the **9** in the number 0.197

.... **(1 mark)**

G **2** Write these numbers in order of size. Start with the smallest number.

5.6 8.7 3.5 1.9 8.3 7

.... **(1 mark)**

F **3** Write these numbers in order of size. Start with the smallest number.

0.541 0.45 0.04 0.5 0.045

.... **(1 mark)**

F **4** Write the following numbers in order of size. Start with the smallest number.

Guided

EXAM ALERT

0.82 0.815 0.8 0.89 0.879

Exam questions similar to this have proved especially tricky – be prepared! ResultsPlus

0.820, 0.815, ~~0.800~~, ~~0.890~~, 0.879

0.8 .... 0.89 **(1 mark)**

D **5** Using the information that $5.8 \times 34 = 197.2$ write down the value of

(a) 58×34

Guided

$58 \times 34 = 197.2 \times$ ....

= ....

5.8 has been multiplied by 10 and 34 hasn't changed. So the answer needs to be multiplied by 10.

(1 mark)

(b) 5.8×3.4

Guided

$5.8 \times 3.4 = 197.2 \div$ ....

= ....

5.8 hasn't been changed and 34 has been divided by 10. So the answer needs to be divided by 10.

(1 mark)

D **6** Using the information that $26 \times 128 = 3328$ write down the value of

(a) 2.6×128

.... **(1 mark)**

(b) 2.6×12.8

.... **(1 mark)**

(c) $3328 \div 2.6$

.... **(1 mark)**

Operations on decimals

 **1** Work out

(a) 25.7 + 7.86

2 5 . 7 0
\+ 7 . 8 6

.... **(1 mark)**

Write in 0s so that all numbers have the same number of decimal places.

(b) 87.2 − 15.48

8 7 . ¹2̸ ¹0
− 1 5 . 4 8

..............2 **(1 mark)**

 **2** Work out

(a) 116 + 23.6 + 5.64

(1 mark)

(b) 78.2 − 6.54

(1 mark)

 **3** Work out

(a) 0.6 × 0.3

Guided

First work out 6 × 3. There are 2 decimal places in total in the calculation so put 2 decimal places in your answer.

(1 mark)

(b) 73.2 ÷ 3

....
3) 7 3 . 2

(1 mark)

 **4** Four friends equally share the cost of a restaurant bill.
The bill is £78.56
How much does each pay?

£..................... **(2 marks)**

E **5** The cost of a calculator is £7.42
Work out the cost of 26 of these calculators.

£..................... **(3 marks)**

D **6** Work out 5.14 × 2.6

First work out 514 × 26. There are 3 decimal places in total in the calculation so put 3 decimal places in your answer.

.... **(3 marks)**

Estimating answers

E **1** Work out an estimate for the value of 19.8×3.4

Guided

$19.8 \times 3.4 \approx 20 \times \ldots\ldots\ldots\ldots$

$= \ldots\ldots\ldots\ldots$ **(2 marks)**

E **2** Work out an estimate for the value of $278 \div 11.3$

$\ldots\ldots\ldots\ldots\ldots\ldots$ **(2 marks)**

D **3** Work out an estimate for the value of $\dfrac{3981}{2.3 \times 18.7}$

Guided

$\dfrac{3981}{2.3 \times 18.7} \approx \dfrac{4000}{2 \times \ldots\ldots\ldots\ldots} = \dfrac{\ldots\ldots\ldots\ldots}{\ldots\ldots\ldots\ldots}$

First round each number to one significant figure.

$= \ldots\ldots\ldots\ldots$ **(2 marks)**

D **4** Work out an estimate for the value of $\dfrac{612 \times 39}{187}$

$\ldots\ldots\ldots\ldots\ldots\ldots$ **(2 marks)**

C **5** Work out an estimate for the value of $\dfrac{40.7 \times 1.6}{0.53}$

Exam questions similar to this have proved especially tricky – be prepared! ResultsPlus

Guided

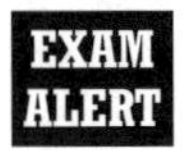

EXAM ALERT

$\dfrac{40.7 \times 1.6}{0.53} \approx \dfrac{40 \times \ldots\ldots\ldots\ldots}{0.5}$

$= \dfrac{\ldots\ldots\ldots\ldots}{0.5}$

$= \dfrac{\ldots\ldots\ldots\ldots}{0.5} \times \dfrac{10}{10}$

$= \dfrac{\ldots\ldots\ldots\ldots}{5}$

$= \ldots\ldots\ldots\ldots$ **(3 marks)**

C **6** Work out an estimate for the value of $\dfrac{9.73 \times 4.12}{0.0214}$

$\ldots\ldots\ldots\ldots\ldots\ldots$ **(3 marks)**

Negative numbers

F **1** (a) Write these numbers in order of size. Start with the smallest number.

−5 4 −6 0 −2

Guided

−5, 4, ~~−6~~, 0, −2

−6 .... 4 **(1 mark)**

(b) Work out

(i) −10 + 4

(ii) 7 − −8

Replace the − − with +.

.... **(1 mark)** .... **(1 mark)**

F **2** (a) Write these numbers in order of size. Start with the smallest number.

2 −8 −5 3 −2 −9

.... **(1 mark)**

(b) Work out

(i) −6 × −3

(ii) 35 ÷ −7

.... **(1 mark)** .... **(1 mark)**

F **3** The table gives information about the temperatures in five cities.

City	London	York	Edinburgh	Leeds	Plymouth
Temperature	2 °C	−1 °C	−6 °C	0 °C	3 °C

(a) Work out the difference between the temperature in York and the temperature in Edinburgh.

......................°C **(1 mark)**

(b) Work out the difference between the temperature in Plymouth and the temperature in Edinburgh.

......................°C **(1 mark)**

(c) The temperature in Aberdeen is 8 °C lower than the temperature in London.
Write down the temperature in Aberdeen.

......................°C **(1 mark)**

F **4** Sasha says, 'The temperature halfway between −6 °C and 4 °C is −2 °C.'
Is Sasha correct? You must give a reason for your answer.

Work out the answer, then compare it with −2

Exam questions similar to this have proved especially tricky – be prepared! ResultsPlus

(2 marks)

Squares, cubes and roots

F **1** Here is a list of numbers.

6 8 10 12 20 36 50

From the numbers in the list, write down

(a) a square number

.... **(1 mark)**

(b) a cube number

.... **(1 mark)**

(c) the square root of 100

.... **(1 mark)**

F **2** Write down

(a) the square of 7

.... **(1 mark)**

(b) the square root of 25

.... **(1 mark)**

(c) the cube of 4

.... **(1 mark)**

(d) the cube root of 8

.... **(1 mark)**

E **3** Work out

(a) 8^2

.... **(1 mark)**

(b) 5^3

.... **(1 mark)**

(c) $\sqrt{81}$

.... **(1 mark)**

(d) $\sqrt[3]{64}$

.... **(1 mark)**

E **4** (a) Work out $3^2 \times 10^3$

Guided

$3^2 \times 10^3 = 9 \times$ .... = ... **(2 marks)**

(b) Write down an estimate for $\sqrt{50}$

.... **(1 mark)**

D **5** Jean says, 'If you add two different square numbers then you always get an odd number.'
Jean is wrong. Explain why.

EXAM ALERT

Give an example to show that Jean is wrong.

....

....

....

Exam questions similar to this have proved especially tricky – be prepared! ResultsPlus

(2 marks)

D **6** Jerry says, 'The cube of 4 is the same as the square of 8'
Is Jerry right? You must explain your answer.

....

....

(2 marks)

Factors, multiples and primes

G 1 Write down all the factors of 24

Guided

1 × 24, 2 × ...., 3 × ...., 4 × ....

The factors of 24 are 1, 2, 3, 4, ..., ..., ..., 24 **(2 marks)**

G 2 Here is a list of eight numbers.

4 7 9 15 18 24 25

From the numbers in the list, write down

(a) a factor of 12

.... **(1 mark)**

(b) a multiple of 5

.... **(1 mark)**

(c) a prime number

.... **(1 mark)**

(d) two even numbers that have a sum of 22

.... and .... **(1 mark)**

F 3 Find the common factors of

(a) 4 and 12 .... **(1 mark)**

(b) 16 and 40 .... **(1 mark)**

(c) 15, 35 and 45 .... **(1 mark)**

C 4 Write 60 as the product of its prime factors.

Guided

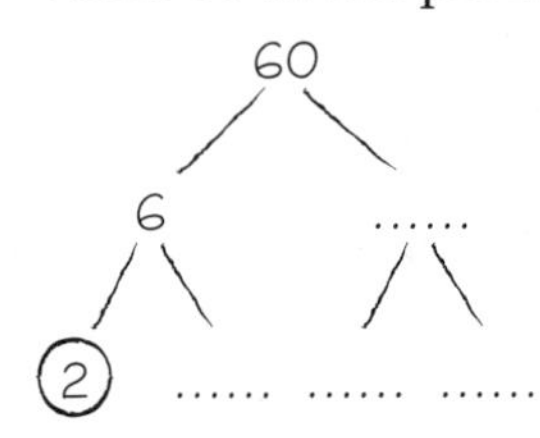

Use a factor tree to find the prime factors. Circle the prime factors as you go along.

'Product' means the prime factors must be multiplied together.

60 = 2 × .... × .... × .... **(2 marks)**

C 5 Write 150 as the product of its prime factors.

150 = ... **(2 marks)**

C 6 Write 72 as the product of its prime factors.

72 = ... **(3 marks)**

HCF and LCM

D **1** Find the lowest common multiple (LCM) of 10 and 12

> Write down a list of multiples of each number and then circle the smallest number that appears in both lists.

Guided

10: 10, 20, 30, ...

12: 12, 24, ... LCM = ... **(2 marks)**

C **2** (a) Express the following numbers as products of their prime factors.

(i) 90

Guided

> Use a factor tree to find the prime factors. Circle the prime factors as you go along.

$90 = 2 \times 3 \times$ .... $\times$ .... **(4 marks)**

(ii) 120

Guided

$120 = 2 \times$ .... $\times$ .... $\times 3 \times$ .... **(4 marks)**

(b) Find the highest common factor (HCF) of 90 and 120

Guided

HCF of 90 and 120 = ... $\times$ .... $\times$ ....

= ....

> Circle the common prime factors from part (a) and then multiply these together.

(2 marks)

C **3** (a) Express 84 as a product of its prime factors.

84 = ... **(2 marks)**

(b) Find the highest common factor (HCF) of 84 and 120

HCF = ... **(2 marks)**

C **4** Find the lowest common multiple (LCM) of 48 and 60

LCM = ... **(2 marks)**

Fractions

 **1** (a) Write $\frac{4}{10}$ in its simplest form.

Guided $\frac{4}{10} = \frac{\dots\dots\dots}{5}$ **(1 mark)**

(b) Work out $\frac{1}{5}$ of £65

Guided $\frac{1}{5}$ of £65 = 65 ÷ 5 = £........... **(1 mark)**

 **2** (a) Write down the fraction of this shape that is shaded.
Give your fraction in its simplest form.

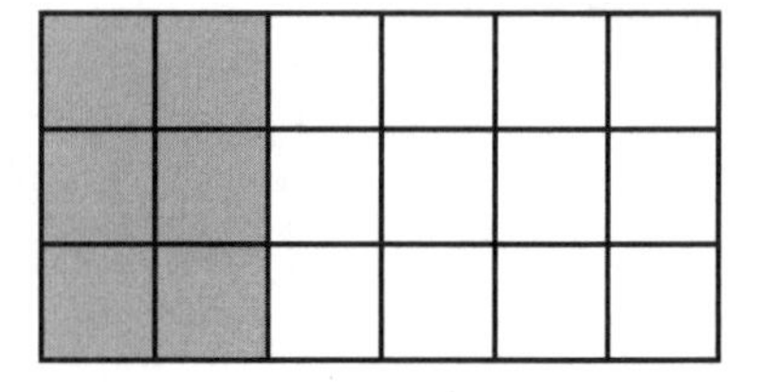

.... **(2 marks)**

(b) Which of these fractions are equivalent to $\frac{1}{2}$?

$\frac{3}{4}$ $\frac{4}{8}$ $\frac{6}{10}$ $\frac{7}{14}$ $\frac{9}{20}$

.... **(2 marks)**

 **3** Work out $\frac{3}{5}$ of 30 kg.

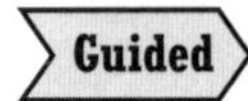 $\frac{1}{5}$ of 30 kg = 30 ÷ .... = ...

$\frac{3}{5}$ of 30 kg = ... × 3

= .... kg **(2 marks)**

 **4** Work out $\frac{2}{9}$ of £180

£..................... **(2 marks)**

E **5** Sally buys a television in a sale.
The normal price of the television is £280
In the sale, all prices are reduced by $\frac{1}{4}$
Work out the price of the television in the sale.

£..................... **(2 marks)**

Simple fractions

1 Work out

(a) $\frac{7}{20} + \frac{1}{5}$

Guided

$\frac{7}{20} + \frac{1}{5} = \frac{7}{20} + \frac{\ldots\ldots}{20}$

$= \frac{\ldots\ldots}{\ldots\ldots}$ **(2 marks)**

(b) $\frac{1}{4} \times \frac{7}{9}$

$\frac{1}{4} \times \frac{7}{9} = \frac{1 \times 7}{\ldots\ldots}$

$= \frac{\ldots\ldots}{\ldots\ldots}$ **(1 mark)**

E **2** Work out

(a) $\frac{8}{9} - \frac{1}{3}$

.... **(2 marks)**

(b) $\frac{5}{11} \div \frac{1}{2}$

$\frac{5}{11} \div \frac{1}{2} = \frac{5}{11} \times \frac{2}{1}$

.... **(2 marks)**

D **3** (a) Work out $\frac{2}{5} + \frac{1}{3}$

.... **(2 marks)**

(b) Work out $\frac{8}{9} \times \frac{3}{5}$

Give your answer in its simplest form.

.... **(2 marks)**

D **4** Some students each chose an activity.
$\frac{2}{3}$ of the students chose climbing.
$\frac{1}{4}$ of the students chose archery.
The rest chose fencing.
What fraction of the students chose fencing?

Guided

$\frac{2}{3} + \frac{1}{4} = \frac{\ldots\ldots}{12} + \frac{\ldots\ldots}{12} = \frac{\ldots\ldots}{\ldots\ldots}$

$1 - \frac{\ldots\ldots}{\ldots\ldots} = \frac{12}{12} - \frac{\ldots\ldots}{\ldots\ldots} = \frac{\ldots\ldots}{\ldots\ldots}$ **(3 marks)**

D **5** A full glass holds $\frac{2}{5}$ of a litre.
How many glasses can be filled from a 2 litre bottle of drink?

.... glasses **(2 marks)**

Mixed numbers

C **1** Work out

$5\frac{1}{4} + 2\frac{2}{5}$

Give your answer as a mixed number in its simplest form.

Guided

$5\frac{1}{4} + 2\frac{2}{5} = \frac{\dots\dots}{4} + \frac{\dots\dots}{\dots\dots}$ Convert the mixed numbers to improper fractions.

$= \frac{\dots\dots}{20} + \frac{\dots\dots}{20}$ Write as equivalent fractions with the same denominator.

$= \frac{\dots\dots}{20}$

$= \dots\dots \frac{\dots\dots}{20}$ Write your final answer as a mixed number in its simplest form.

(3 marks)

C **2** Work out (a) $6\frac{5}{7} + 4\frac{1}{2}$ (b) $4\frac{1}{2} - \frac{7}{8}$

Give each answer as a mixed number in its simplest form.

(a) .... **(3 marks)** (b) .... **(3 marks)**

C **3** Work out $4\frac{3}{4} \div 3\frac{1}{2}$

Give your answer as a mixed number in its simplest form.

Exam questions similar to this have proved especially tricky – be prepared! ResultsPlus

Guided

EXAM ALERT

$4\frac{3}{4} \div 3\frac{1}{2} = \frac{\dots\dots}{4} \div \frac{\dots\dots}{2}$ Write both numbers as improper fractions.

$= \frac{\dots\dots}{4} \times \frac{2}{\dots\dots}$ Change ÷ to × and turn second fraction upside down.

$= \frac{\dots\dots}{\dots\dots}$

$= \dots\dots \frac{\dots\dots}{\dots\dots}$ Write your final answer as a mixed number in its simplest form.

(3 marks)

C **4** Work out (a) $2\frac{1}{2} \times \frac{3}{10}$ (b) $8\frac{1}{3} \div 3\frac{3}{4}$

Give each answer as a mixed number in its simplest form where appropriate.

(a) .... **(3 marks)** (b) .... **(3 marks)**

Number and calculator skills

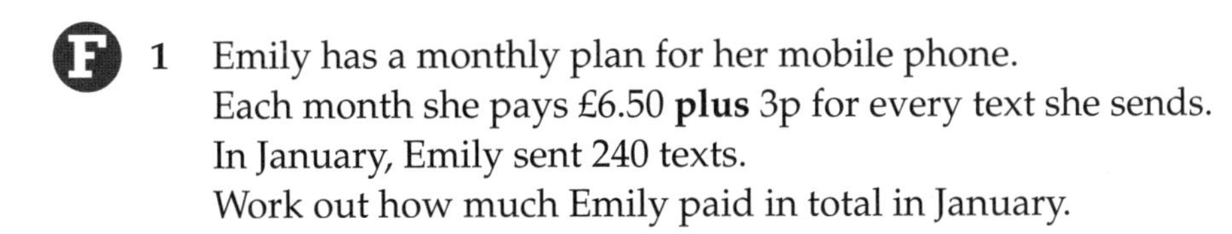

1 Emily has a monthly plan for her mobile phone.
Each month she pays £6.50 **plus** 3p for every text she sends.
In January, Emily sent 240 texts.
Work out how much Emily paid in total in January.

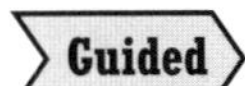

Amount spent on texts = ... × 3
= ....pence
= £...........
Total amount spent = £6.50 + ...
= £...........

(3 marks)

2 Karim buys:

- 3 calculators costing £7.50 each
- 4 pencils costing 79p each.

He pays with two £20 notes.
How much change should Karim get?

£.................... **(4 marks)**

3 Work out $20 - 2 \times 3$

$20 - 2 \times 3 = 20 -$ ....
= ....

Use **BIDMAS** to get the correct order of operations.

Exam questions similar to this have proved especially tricky – be prepared!

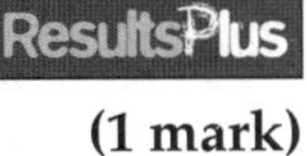

(1 mark)

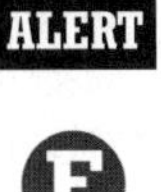

4 Work out $6 \times 3 + 24 \div 4$

.... **(2 marks)**

5 Use your calculator to work out the value of $5.6^2 + 2 \times 8.5$
Write down all the figures on your calculator display.

.... **(2 marks)**

6 Use your calculator to work out the value of $\sqrt{47.8^2 - 567}$
Write down all the figures on your calculator display.
Give your answer as a decimal.

.... **(2 marks)**

7 Use your calculator to work out the value of $\dfrac{\sqrt{13.5 + 3.4^2}}{2.3 \times 1.5}$

Write down all the figures on your calculator display.

.... **(3 marks)**

Percentages

G **1** (a) Write 47% as a fraction.

Guided

$47\% = \frac{47}{\dots\dots}$

(1 mark)

(b) Write 30% as a fraction in its simplest form.

Guided

$30\% = \frac{30}{\dots\dots} = \frac{\dots\dots}{\dots\dots}$

(2 marks)

F **2** Work out 40% of £320

Guided

40% of £320 = $\frac{40}{\dots\dots}$ × .... = £...........

(2 marks)

F **3** (a) Work out 20% of 75 m.

.... m **(2 marks)**

(b) Work out 32% of £250

£..................... **(2 marks)**

D **4** Uzma invests £4000 in a bank account for 1 year.
Interest is paid at a rate of 2.5% per annum.
How much interest will Uzma get at the end of 1 year?

Guided

$\frac{\dots\dots}{100}$ × £4000 = £..................... **(2 marks)**

D **5** A farmer has 48 llamas.
30 of the llamas are female.

(a) Write 30 out of 48 as a percentage.

.... **(2 marks)**

30% of the female llamas are pregnant.

(b) Work out 30% of 30

.... **(2 marks)**

C **6** At an outdoor centre, 140 students each choose one activity.
$\frac{1}{7}$ of the students choose rock climbing.
$\frac{3}{7}$ of the students choose rafting.
All the rest of these students choose abseiling.
How many students choose abseiling?

.... **(3 marks)**

Percentage change

1 Helen buys a jacket in a sale.
The normal price of the jacket is reduced by 35%.
The normal price is £84
Work out the sale price of the jacket.

Exam questions similar to this have proved especially tricky – be prepared! ResultsPlus

Reduction $= \frac{35}{100} \times$ ....

= ....

Sale price = 84 − ....

= £...........

(3 marks)

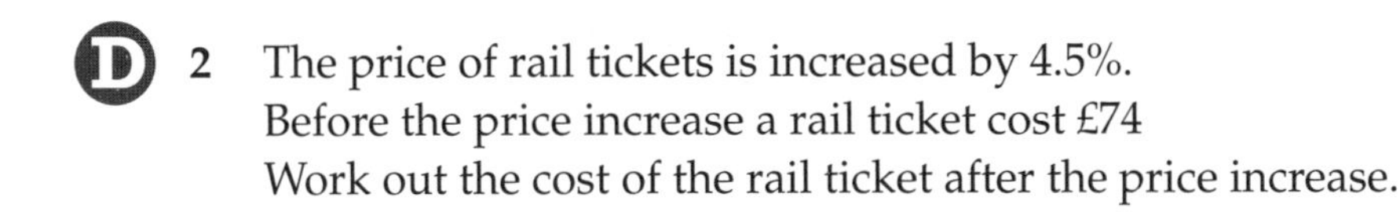

2 The price of rail tickets is increased by 4.5%.
Before the price increase a rail ticket cost £74
Work out the cost of the rail ticket after the price increase.

£.................... **(3 marks)**

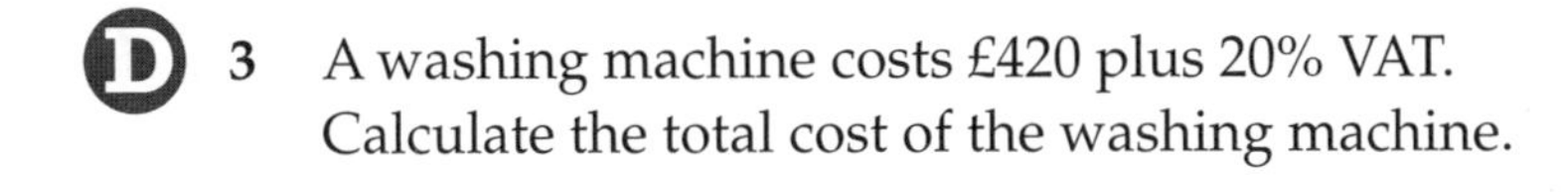

3 A washing machine costs £420 plus 20% VAT.
Calculate the total cost of the washing machine.

£.................... **(3 marks)**

4 Kevin works in a bookshop.
He is paid £156 per week **plus** 8% of the total value of the books he sells that week.
In one week he sold books with a total value of £1200
Work out the total amount Kevin was paid that week.

Money from book sales $= \frac{8}{100} \times$ ....

= £...........

Total amount = 156 + ...

= £...........

(3 marks)

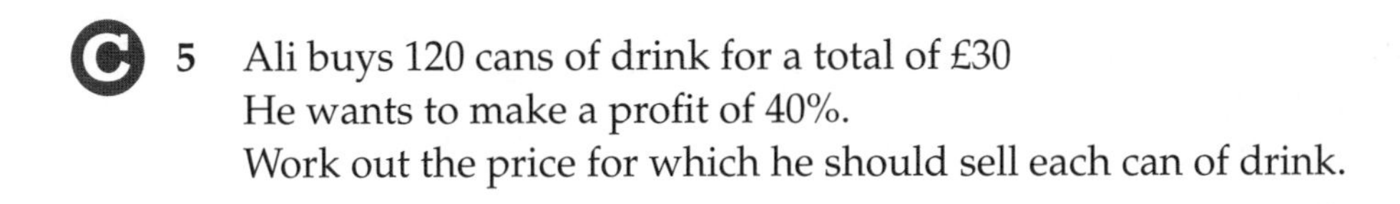

5 Ali buys 120 cans of drink for a total of £30
He wants to make a profit of 40%.
Work out the price for which he should sell each can of drink.

.... p **(4 marks)**

Fractions, decimals and percentages

E **1** Write these numbers in order of size. Start with the smallest number.

$\frac{1}{2}$ 55% $\frac{2}{5}$ 47% $\frac{9}{20}$

Guided $\frac{1}{2} = \frac{\ldots\ldots\ldots\ldots}{100} = \ldots\ldots\ldots\ldots\%$ $\frac{2}{5} = \frac{\ldots\ldots\ldots\ldots}{100} = \ldots\ldots\ldots\ldots\%$ $\frac{9}{20} = \frac{\ldots\ldots\ldots\ldots}{100} = \ldots\ldots\ldots\ldots\%$

Write each number in its original form.

In order the numbers are ...., ..., ..., ..., ... **(2 marks)**

E **2** Write these numbers in order of size. Start with the smallest number.

76% $\frac{3}{4}$ $\frac{7}{10}$ 72% $\frac{13}{20}$

.... **(2 marks)**

D **3** Liam's annual income is £16 000

EXAM ALERT

He pays $\frac{1}{5}$ of the £16 000 in rent.

He spends 15% of the £16 000 on food.

Work out how much of the £16 000 Liam has left.

Exam questions similar to this have proved especially tricky – be prepared! ResultsPlus

Guided $\frac{1}{5}$ of 16 000 = 16 000 ÷ 5

= ....

10% of 16 000 = 16 000 ÷ 10 = ...

5% of 16 000 = ... ÷ 2 = ...

15% of 16 000 = ... + ... = ...

Total amount spent on rent and food = ... + ...

= ....

Amount of money left = 16 000 − ....

= £............ **(4 marks)**

D **4** Lily, Jodie and Tomas work in a restaurant.

They share the tips depending on the amount of time they work.

Lily gets $\frac{1}{2}$ of the tips.

Jodie gets 30% of the tips.

Tomas gets the rest of the tips.

One evening, Tomas got a total of £16

How much did Lily get?

£.................... **(4 marks)**

Ratio

E **1** Write the ratio 15 : 25 in its simplest form.

Divide both sides of the ratio by 5.

.... **(1 mark)**

D **2** There are 60 toy cars in a box.
18 of the toy cars are blue. The rest of the toy cars are red.
Write down the ratio of the number of red toy cars to the number of blue toy cars.
Give your ratio in its simplest form.

Guided

Number of red cars = 60 − ....

= ....

Make sure you put the numbers in the ratio in the correct order.

Ratio of red cars to blue cars = ... : ....

= .... : .... **(2 marks)**

D **3** There are 32 students in a class. 20 of the students are girls.
Rosie says, 'The ratio of the number of girls to the number of boys in this class is 3 : 5'
Is Rosie right?
You must give a reason for your answer.

....

(2 marks)

D **4** Ahmed and James share £120 in the ratio 1 : 3
How much does James get?

Exam questions similar to this have proved especially tricky – be prepared! ResultsPlus

Guided

EXAM ALERT

Number of shares = 1 + 3 = ...

One share is worth £120 ÷ .... = £...........

James gets .... × .... = £........... **(2 marks)**

C **5** Linda, Mel and Tomos share the driving on a journey in the ratio 2 : 3 : 4
Mel drove a distance of 240 km.
Work out the length of the journey.

.... km **(2 marks)**

C **6** A 200 g jar of coffee costs £3.90
A 150 g jar of coffee costs £2.85
Which size jar is better value for money?
You must show your working.

Work out how much 1 g of coffee costs for each size.

.... **(3 marks)**

Problem-solving practice

1 Ben buys

1 newspaper costing £1.25
2 bars of chocolate costing 73p each
and 1 magazine

He pays with a £10 note. He gets £4.84 change.
How much did the magazine cost?

Work out the cost of the newspaper and **two** bars of chocolate. Then add the amount of change to your answer. Now subtract your total from £10

.... **(4 marks)**

2 A football team is playing an away match. 230 supporters are going to the match.
The manager of the football club is going to book some coaches.
Each coach will hold a maximum of 52 people.
How many coaches does the manager need to book?

Number of coaches needed = 230 ÷ ....
= .... (to 1 d.p)
= .... coaches

Give a whole number of coaches for your answer.

(3 marks)

3 Mrs Jones is going to take her family to a farm.
She wants to buy tickets for 2 adults and 3 children aged 4, 6 and 9
She pays with two £20 notes. How much change should she get?

Ticket prices (per person)	
Adult	£8.50
Child (5–14)	£5.50
Child (under 5)	free

Carefully note the ages of the children and work out the total cost of the tickets. Then work out Mrs Jones's change.

£.................... **(4 marks)**

4 Sarah is going to walk from her house to the cinema.
It will take Sarah 20 minutes to walk to the cinema.
She wants to meet her friend at the cinema 15 minutes before the film starts.
Sarah's film starts at 19:20
At what time should Sarah leave her house?

Work backwards from 19:20

.... **(3 marks)**

Problem-solving practice

*5 Tickets R-US and Cheap Tickets both advertise tickets for the same concert.

The question has a * next to it, so make sure that you show all your working and write your answer clearly in the context of the problem.

Tickets R-US
£36 plus 5% booking fee

Cheap Tickets
£35 plus 7.5% booking fee

Helen wants to pay the least money possible for a ticket.
Which shop should she buy her ticket from, Tickets R-US or Cheap Tickets?

Work out the price plus the booking fee for each ticket.

(4 marks)

6 Last year, Kevin spent

$\frac{1}{8}$ of his salary on entertainment
$\frac{2}{5}$ of his salary on rent
15% of his salary on living expenses.

He saved the rest of his salary.
Last year Kevin's salary was £32 000
How much money did Kevin save?

Amount spent on entertainment = 32 000 ÷ ....
= ....

You should show **all** your working.

Amount spent on rent = 32 000 ÷ .... × ....
= ....

10% of 32 000 = ...

5% of 32 000 = ...

15% of 32 000 = ...

Amount spent on living expenses = £...............

Total amount spent = ... + ... + ...
= £.............

Amount of money saved = 32 000 − ....
= £.............

(4 marks)

Collecting like terms

E **1** Simplify

(a) $t + t + t$

Guided $t + t + t = ...t$

(1 mark)

(b) $5bc - 2bc$

$5bc - 2bc = ...bc$

(1 mark)

(c) $4k + 5m + 7k - 2m$

Guided $4k + 5m + 7k - 2m = 4k + 7k + 5m - 2m$

$= ....k + ...m$ **(2 marks)**

E **2** Simplify

(a) $m + p + m + m + m + p$

.... **(1 mark)**

(b) $7x - 3x$

.... **(1 mark)**

(c) $p^2 + p^2$

.... **(1 mark)**

E **3** (a) $5a + 4b + 6a + b$

.... **(2 marks)**

(b) $8e + 6f - 2e + f$

.... **(2 marks)**

E **4** Alex says that $6x - 3 = 5$ is an expression.
Sam says that $6x - 3 = 5$ is an equation.
Who is right? Explain why.

....

....

(2 marks)

E **5** Simplify $5x - 2y - 3x - 8y$

Guided $5x - 2y - 3x - 8y = 5x - 3x - 2y - 8y$

$= ....x - ....y$ **(2 marks)**

E **6** Simplify

(a) $10 + 2a - 3 + 5a$

.... **(2 marks)**

(b) $9p - 3t - 5p - 2t$

.... **(2 marks)**

E **7** Simplify

(a) $3xy + 4xy - xy$

.... **(1 mark)**

(b) $3r + 6t - r + 2t + 4$

.... **(2 marks)**

Simplifying expressions

D **1** Simplify

(a) $r \times r \times r$

Guided

$r \times r \times r = r^{\dots\dots}$

(1 mark)

(b) $3m \times p$

$3m \times p = \dots\dots\dots\dots mp$

(1 mark)

(c) $4x \times 5y$

Guided

$4x \times 5y = \dots\dots\dots\dots xy$

(1 mark)

D **2** Simplify

(a) $a \times a \times a \times a \times a$

.... **(1 mark)**

(b) $k \times 8n$

.... **(1 mark)**

(c) $6h \times 3j$

.... **(1 mark)**

D **3** Simplify

(a) $e \times e \times f \times f \times f$

.... **(1 mark)**

(b) $8x \times 4y$

.... **(1 mark)**

D **4** Simplify

(a) $24a \div 8$

Guided

$24a \div 8 = \frac{24a}{8}$

$= \dots\dots\dots\dots a$ **(1 mark)**

(b) $30xy \div 5x$

$30xy \div 5x = \frac{30xy}{5x}$

$= \dots\dots\dots\dots$ **(2 marks)**

D **5** Simplify

(a) $18b \div 6$

.... **(1 mark)**

(b) $56mp \div 8m$

.... **(1 mark)**

D **6** Simplify

(a) $7g \times 4h$

.... **(1 mark)**

(b) $32tz \div 4$

.... **(1 mark)**

(c) $48mn \div 6n$

.... **(1 mark)**

(d) $4a \times 2b \times c$

.... **(1 mark)**

Indices

1 Simplify

(a) $m^3 \times m^9$

Guided

$m^3 \times m^9 = m^{3+9}$

$= m^{\cdots\cdots}$ **(1 mark)**

(b) $p^{10} \div p^2$

$p^{10} \div p^2 = p^{10-2}$

$= p^{\cdots\cdots}$ **(1 mark)**

(c) $(t^4)^5$

Guided

$(t^4)^5 = t^{4 \times 5}$

$= t^{\cdots\cdots}$ **(1 mark)**

2 Simplify

(a) $d^8 \times d^5$

.... **(1 mark)**

(b) $e^6 \div e^2$

.... **(1 mark)**

(c) $(f^2)^4$

.... **(1 mark)**

3 Simplify

Multiply or divide any number parts first.

(a) $2g^3 \times 7g^6$

Guided

$2g^3 \times 7g^6 = 2 \times 7 \times g^{\cdots\cdots + \cdots\cdots}$

$= \ldots\ldots\ldots\ldots g^{\cdots\cdots}$

(1 mark)

(b) $12b^7 \div 3b^4$

$12b^7 \div 3b^4 = 12 \div \ldots\ldots\ldots \times b^{\cdots\cdots - \cdots\cdots}$

$= \ldots\ldots\ldots\ldots b^{\cdots\cdots}$

(1 mark)

4 Simplify

(a) $g \times g^6$

$g = g^1$

.... **(1 mark)**

(b) $k^9 \div k^3 \times k^2$

.... **(1 mark)**

5 Simplify

(a) $\frac{x^5 \times x^4}{x^6}$

Guided

$\frac{x^5 \times x^4}{x^6} = \frac{x^{\cdots\cdots}}{x^6}$

$= x^{\cdots\cdots}$ **(2 marks)**

(b) $\frac{y^{12}}{y^3 \times y}$

$\frac{y^{12}}{y^3 \times y} = \frac{y^{12}}{y^{\cdots\cdots}}$

$= y^{\cdots\cdots}$ **(2 marks)**

6 Simplify

(a) $\frac{h^4 \times h}{h^3}$

.... **(2 marks)**

(b) $\frac{a^{15}}{a^2 \times a^7}$

.... **(2 marks)**

Expanding brackets

1 Expand

(a) $4(a - 2b)$

Guided: $4(a - 2b) = \ldots\ldots a - \ldots\ldots b$ **(1 mark)**

(b) $d(d + 5)$

$d(d + 5) = d^{\cdots\cdots} + \ldots\ldots$ **(1 mark)**

2 Expand

(a) $3(5b + 1)$

.... **(1 mark)**

(b) $e(e - 2)$

.... **(1 mark)**

3 Expand

(a) $2d(d - 3)$

.... **(1 mark)**

(b) $3f(1 - e)$

.... **(1 mark)**

4 Expand

(a) $6(3 - 4d)$

.... **(1 mark)**

(b) $p^2(3p - 1)$

.... **(1 mark)**

5 Expand and simplify $3(2x + 5y) + 4(x - 2y)$

$3(2x + 5y) + 4(x - 2y) = 6x + 15y + \ldots\ldots\ldots\ldots x - \ldots\ldots\ldots\ldots y$

$= \ldots\ldots\ldots\ldots x + \ldots\ldots\ldots\ldots y$ **(2 marks)**

6 Expand and simplify

(a) $2(k + 3) + 4(k + 1)$

.... **(2 marks)**

(b) $7(2x - y) - 3(3x - 2y)$

Each number in the second bracket is multiplied by –3.

.... **(2 marks)**

7 Expand and simplify $2(3 - x) - 6(1 - 2x)$

.... **(2 marks)**

Factorising

D **1** Factorise

EXAM ALERT

Exam questions similar to this have proved especially tricky – be prepared! ResultsPlus

(a) $3x + 18$

Guided

$3x + 18 = 3(\ldots\ldots\ldots\ldots + \ldots\ldots\ldots\ldots)$ **(1 mark)**

(b) $y^2 - 9y$

Guided

$y^2 - 9y = y(\ldots\ldots\ldots\ldots - \ldots\ldots\ldots\ldots)$ **(1 mark)**

D **2** Factorise

(a) $5m + 20$

.... **(1 mark)**

(b) $8t - 14$

.... **(1 mark)**

(c) $v^2 - v$

.... **(1 mark)**

D **3** Factorise

(a) $9 - 3k$

.... **(1 mark)**

(b) $2m^2 - m$

.... **(1 mark)**

D **4** Factorise fully

'Factorise fully' means that you need to take out the highest common factor.

(a) $16p - 24$

Guided

$16p - 24 = 8(\ldots\ldots\ldots\ldots - \ldots\ldots\ldots\ldots)$ **(1 mark)**

(b) $4x^2 + 8x$

Guided

$4x^2 + 8x = 4x(\ldots\ldots\ldots\ldots + \ldots\ldots\ldots\ldots)$ **(1 mark)**

C **5** Factorise fully $8mp - 12m^2$

.... **(2 marks)**

C **6** Factorise fully

(a) $14ab - 21bc$

.... **(2 marks)**

(b) $8x^2 + 10xy$

.... **(2 marks)**

C **7** Factorise fully

(a) $12xy - 16x$

.... **(2 marks)**

(b) $20y - 40y^2$

.... **(2 marks)**

Sequences

F **1** Here are the first four terms in a number sequence.

3 7 11 15

(a) Write down the next **two** terms in this number sequence.

.... **(2 marks)**

(b) Explain how you found your answer.

Write down the rule that you used.

....

(2 marks)

(c) Write down the ninth term in this number sequence.

Carry on the sequence up to the ninth term.

.... **(2 marks)**

F **2** Here are the first four terms of a number sequence.

78 72 66 60

(a) Write down the next **two** terms of this number sequence.

.... **(2 marks)**

(b) Zach says that 40 is a number in this sequence.
Is Zach right? You must give a reason for your answer.

....

....

(2 marks)

E **3** Here are some patterns made using sticks.

Pattern number 1 Pattern number 2 Pattern number 3

(a) In the space below draw pattern number 4

Copy pattern number 3 and then draw in some more sticks to make pattern number 4.

(1 mark)

(b) Complete the table.

Pattern number	1	2	3	4	5
Number of sticks	5	9	13		

(2 marks)

(c) Find the number of sticks used for pattern number 9

.... **(2 marks)**

(d) Clare says that there is a pattern number that will use 100 sticks.
Is Clare right? You must explain your answer.

....

....

(1 mark)

*n*th term of a sequence

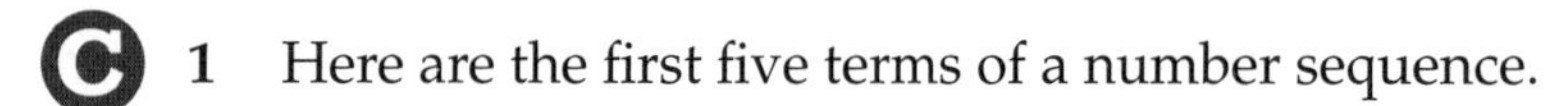

1 Here are the first five terms of a number sequence.

1 5 9 13 17

Find an expression, in terms of n, for the nth term of the sequence.

Guided

zero term + ... + ... + ... + ...

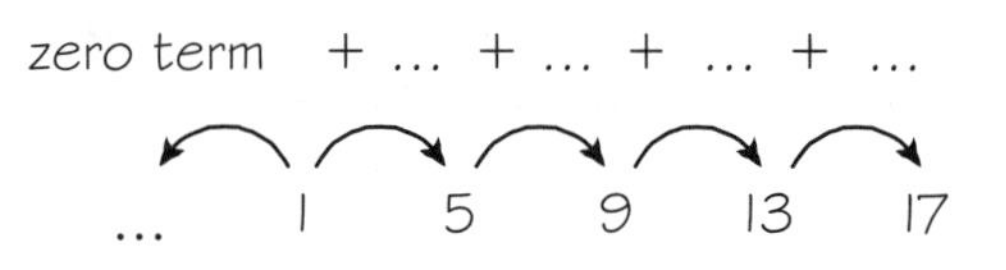

Work out the difference between each term. Then work out the zero term.

nth term = ...n + ... = ...

(2 marks)

2 Here are the first five terms of a number sequence.

17 12 7 2 −3

Find an expression, in terms of n, for the nth term of the sequence.

.... **(2 marks)**

3 (a) Here are the first five terms of a number sequence.

2 6 10 14 18

Find an expression, in terms of n, for the nth term of the sequence.

.... **(2 marks)**

(b) Paul says that 73 is a term in this sequence. Paul is wrong. Explain why.

Look at the type of numbers in the sequence.

....

(1 mark)

4 (a) The nth term of a sequence is $8n + 3$
Write down the first three terms of this sequence.

Guided

1st term $n = 1$ $8 \times 1 + 3 =$ ....

2nd term $n =$ .... $8 \times$ .... $+ 3 =$ ....

3rd term $n =$ .... $8 \times$ .... $+ 3 =$ .... **(2 marks)**

(b) Jenny says that 45 is a term in this sequence. Jenny is wrong. Explain why.

Try and find a value for n that gives a result of 45.

.... **(1 mark)**

5 The nth term of a sequence is $3n - 1$
Work out the 50th term of this sequence.

.... **(1 mark)**

Equations 1

F **1** Solve

(a) $4k = 20$

Guided

$4k = 20 \quad (\div 4)$

$k = \ldots\ldots\ldots\ldots$ **(1 mark)**

(b) $m - 6 = 13$

$m - 6 = 13 \quad (+ 6)$

$m = \ldots\ldots\ldots\ldots$ **(1 mark)**

F **2** Solve

(a) $p + 9 = 12$

$p = \ldots\ldots\ldots\ldots$ **(1 mark)**

(b) $\frac{x}{5} = 7$

$x = \ldots\ldots\ldots\ldots$ **(1 mark)**

F **3** Solve

(a) $x + x + x = 18$

$x = \ldots\ldots\ldots\ldots$ **(1 mark)**

(b) $2y = 5$

Give the answer as a mixed number or a decimal.

$y = \ldots\ldots\ldots\ldots$ **(1 mark)**

4 Solve

(a) $20 - h = 12$

$h = \ldots\ldots\ldots\ldots$ **(1 mark)**

(b) $3m + 2m = 30$

$m = \ldots\ldots\ldots\ldots$ **(1 mark)**

5 Solve $2x - 3 = 9$

Exam questions similar to this have proved especially tricky – be prepared! ResultsPlus

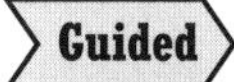

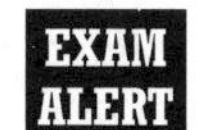

$2x - 3 = 9 \quad (+ 3)$

$2x = \ldots\ldots\ldots\ldots \quad (\div 2)$

$x = \ldots\ldots\ldots\ldots$

(2 marks)

6 Solve

(a) $5y - 4 = 16$

$y = \ldots\ldots\ldots\ldots$ **(2 marks)**

(b) $3x + 4 = 10$

$x = \ldots\ldots\ldots\ldots$ **(2 marks)**

D **7** Solve $12 = \frac{2x}{3} - 4$

Start by adding 4 to both sides of the equation.

$x = \ldots\ldots\ldots\ldots\ldots\ldots\ldots\ldots$ **(2 marks)**

Equations 2

D **1** Solve

(a) $8x - 3 = 13$

Guided

$8x - 3 = 13$ (+ 3)

$8x =$ .... (÷ 8)

$x =$ .... **(2 marks)**

(b) $4(x - 3) = 16$

$4(x - 3) = 16$

$4x -$ .... $= 16$

$4x =$ ....

$x =$ .... **(3 marks)**

D **2** Solve

(a) $7x + 5 = 33$

$x =$........... **(2 marks)**

(b) $3(2x + 9) = 30$

$x =$........... **(3 marks)**

D **3** Solve $7x - 4 = 3x - 5$

Guided

$7x - 4 = 3x - 5$

...........$x - 4 = -5$

.... $x =$ ....

$x = \frac{...........}{...........}$ **(3 marks)**

D **4** Solve

(a) $6x - 5 = 2x + 8$

$x =$........... **(3 marks)**

(b) $4(x + 3) = 3x - 5$

$x =$........... **(3 marks)**

C **5** Solve $4(2x - 1) = 3(x + 4)$

First multiply out the brackets.

$x =$ .... **(3 marks)**

C **6** Solve $6 - 2x = 3(1 - x)$

$x =$ .... **(2 marks)**

Writing equations

C **1** All the angles in the quadrilateral are measured in degrees.
Work out the size of the largest angle.

> Use the fact that angles in a quadrilateral add up to 360°.

$x + 30°$, $2x$, $x + 80°$, $x - 50°$

Diagram **NOT** accurately drawn

Guided

$x + 30 + 2x + x + 80 + x - 50 = 360$

............x + ... $= 360$

............x = ...

x = ...

$2x = 2 \times$ .... = ...

$x + 80$ = ... $+ 80$ = ...

The largest angle is ....° **(4 marks)**

C **2** All the angles in the triangle are measured in degrees.
Work out the size of the smallest angle.

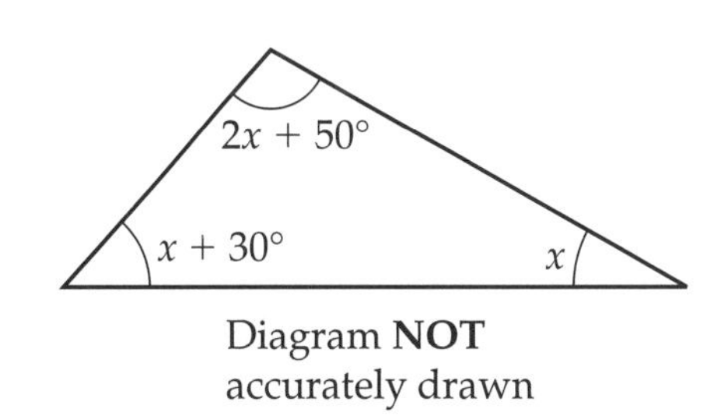

Diagram **NOT** accurately drawn

.....................° **(3 marks)**

C **3** All the lengths on the quadrilateral are measured in centimetres.
The perimeter of the quadrilateral is 39 cm.
Work out the length of the shortest side of the quadrilateral.

> Use the fact that the perimeter is 39 cm to write down an equation. Solve your equation to find x. Use your value of x to work out the value of $x + 3$.

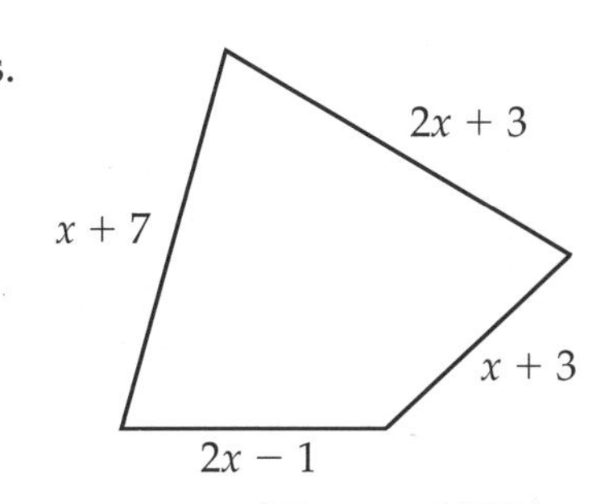

Diagram **NOT** accurately drawn

.... cm **(4 marks)**

Trial and improvement

C 1 The equation $x^3 + 3x = 21$ has a solution between 2 and 3
Use a trial and improvement method to find this solution.
Give your answer correct to 1 decimal place.
You must show **all** your working.

Guided

x	$x^3 + 3x$	Too big or too small
2.5	23.125	too big
2.3	....	....
2.4	....	....
....	....	....

$x =$ .... **(4 marks)**

C 2 The equation $x^3 + 2x^2 = 75$ has a solution between 3 and 4
Use a trial and improvement method to find this solution.
Give your answer correct to 1 decimal place.
You must show **all** your working.

$x =$ .... **(4 marks)**

C 3 The equation $x^3 - 7x = 45$ has a solution between 4 and 5
Use a trial and improvement method to find this solution.
Give your answer correct to 1 decimal place.
You must show **all** your working.

$x =$ .... **(4 marks)**

Inequalities

C **1** An inequality is shown on the number line.

> An 'open' circle means that the inequality is 'greater than' or 'less than'.

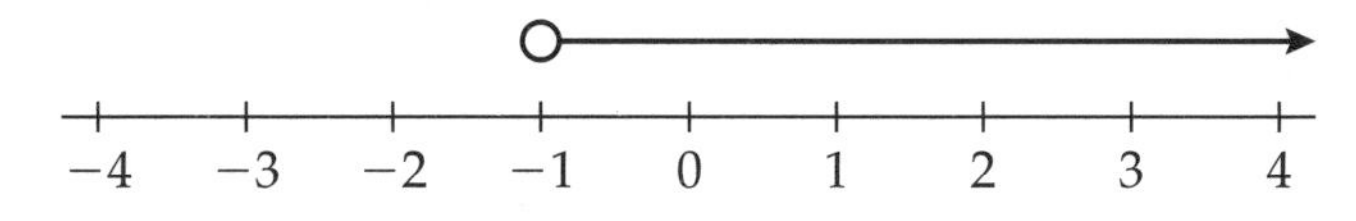

Write down the inequality.

Guided The inequality is x .... -1 **(2 marks)**

C **2** An inequality is shown on the number line.

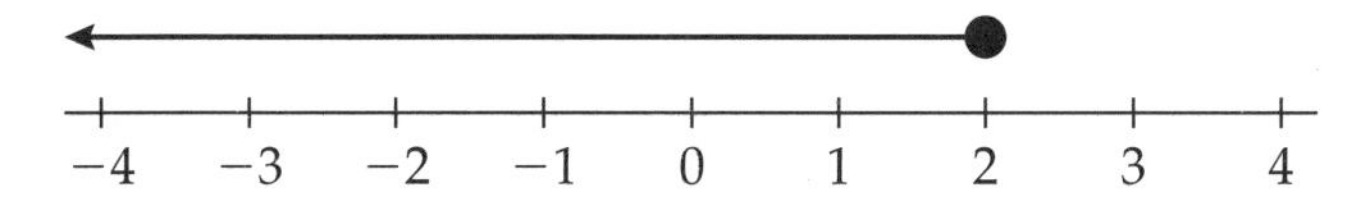

> Use the signs $\leqslant$ or $\geqslant$ when there is a 'closed' circle.

Write down the inequality.

.... **(2 marks)**

C **3** $-3 < n \leqslant 2$
n is an integer.
Write down all the possible values of n.

> The word integer means 'a whole number'.

Guided n could be -2, ... , ... , ... , 2 **(2 marks)**

C **4** $-5 \leqslant m < 1$
m is an integer.
Write down all the possible values of m.

.... **(2 marks)**

C **5** Show the inequality $-1 < x \leqslant 3$ on the number line below.

> There should be an open circle above -1 and a closed circle above 3.

(2 marks)

C **6** Show the inequality $0 \leqslant x < 3$ on the number line below.

(2 marks)

Solving inequalities

1 Solve the inequality $4x - 3 > 21$

Use the inequality symbol throughout.

Guided

$4x - 3 > 21$ $(+ 3)$

$4x >$ .... $(\div 4)$

$x >$ ....

(2 marks)

2 (a) Solve the inequality $5x - 2 < 28$

.... **(2 marks)**

(b) x is a whole number.
Write down the smallest value of x that satisfies $5x - 2 < 28$

Write down the smallest value of x that satisfies your answer to part (a).

.... **(1 mark)**

3 (a) Solve the inequality $6x + 5 < 17$

.... **(2 marks)**

(b) Represent your answer to part (a) on the number line below.

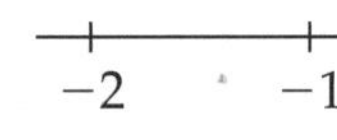

−2 −1 0 1 2 3 4 5

(2 marks)

4 Solve the inequality $3(x - 4) \geqslant 2$

Start by multiplying out the brackets.

.... **(3 marks)**

5 Solve the inequality $2(3x + 1) < 5$

.... **(3 marks)**

Substitution

1 (a) Work out the value of $3p + 2$ when $p = 5$

Guided

$3p + 2 = 3 \times$ .... $+ 2$

$=$ .... $+ 2$

$=$ ....

Use **BIDMAS** to get the correct order of operations.

(2 marks)

(b) Work out the value of $2a + 3b$ when $a = 4$ and $b = 7$

Guided

$2a + 3b = 2 \times$ .... $+ 3 \times$ ....

$=$ .... $+$ ....

$=$ ....

Substitute the correct value for each letter.

(2 marks)

E **2** (a) Work out the value of $8d + 3$ when $d = 3$

.... **(2 marks)**

(b) Work out the value of $5m - 2n$ when $m = 6$ and $n = 3$

.... **(2 marks)**

D **3** (a) Work out the value of $3g^2$ when $g = 4$

Guided

$2g^2 = 3 \times$ ....^{2}

$= 3 \times$ ....

$=$ ....

Substitute 4 for g.

(1 mark)

(b) Work out the value of $4x^2 - 3y$ when $x = -5$ and $y = 20$

Guided

$4x^2 - 3y = 4 \times ($...........$)^2 - 3 \times$ ....

$= 4 \times$ .... $-$ ....

$=$ .... $-$ ....

$(-5)^2 = 5^2$

$=$ ....

(2 marks)

D **4** Work out the value of $5a^2 - 2a$ when $a = -6$

.... **(2 marks)**

D **5** (a) Work out the value of $3(2y - 3t)$ when $y = 4$ and $t = -5$

.... **(2 marks)**

(b) Work out the value of $\frac{p(t - 4)}{2}$ when $p = 3$ and $t = -4$

.... **(2 marks)**

Formulae

F **1** This formula can be used to work out employee pay.

Pay = number of hours worked × rate of pay per hour

Nimer worked for 12 hours.
His rate of pay per hour was £5.50
How much did Nimer earn?

Guided

Nimer's pay = number of hours worked × rate of pay

= .... × ....

= £............

Use the information given in the question.

(2 marks)

2 You can use this formula to work out how long to cook a chicken.

Cooking time in minutes = 45 × weight in kilograms + 20

A chicken weighs 2 kg.
Work out the cooking time for this chicken.

Use **BIDMAS** to get the correct order of operations.

.... minutes **(2 marks)**

3 A plumber uses this formula to work out the amount, P in pounds, he charges for a job.

$P = 30n + 25$

where n is the number of hours he works on the job.
Work out how much the plumber charges for a job lasting 6 hours.

£..................... **(2 marks)**

4 $A = 3t^2 - bc$
Work out the value of A when $b = 3$, $c = 6$ and $t = -5$

Guided

$A = 3 \times (-5)^2 - 3 \times 6$

= 3 × .... − ....

= .... − ....

= .... **(2 marks)**

D **5** Find the value of $4x^2 - 5x$ when $x = -4$

.... **(2 marks)**

Writing formulae

1 James packs books into boxes.
He can fit 30 books into each box.
Write down a formula for the total number of books, T, that he can fit into b boxes.

Total number of books = 30 × number of boxes

.... = 30 × ....

Replace the words by the letters given in the question.

(2 marks)

2 Kelly packs eggs into boxes.
She can fit 12 eggs in each box.
Write down a formula for the total number of eggs, E, that she can fit into x boxes.

.... **(2 marks)**

3 Lesley hires a car.
The cost of hiring a car is £45 plus £30 for each day.
Write down a formula for the total cost, £C, to hire the car for t days.

Total cost = 45 + 30 × number of days

.... = 45 + 30............

(2 marks)

4 A shop sells cakes and buns.
Cakes cost c pence each, buns cost b pence each.
Malik buys 6 cakes and 4 buns.
The total cost is C pence.
Write down a formula for C in terms of c and b.

.... **(3 marks)**

5 The diagram shows a quadrilateral.
Write down a formula, in terms of x, for the perimeter, P, of the rectangle.
Give your answer in its simplest form.

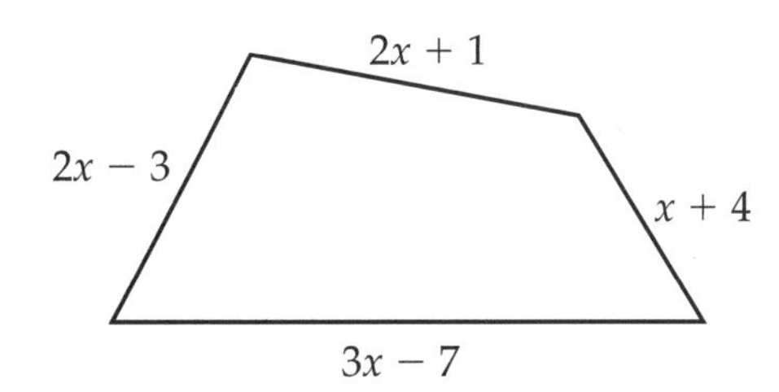

The perimeter is the distance all the way around the shape.

.... **(2 marks)**

6 A bag of apples costs a pence.
A bag of pears costs p pence.
Sue buys 4 bags of apples and 3 bags of pears.
The total cost is £C.
Write down a formula for C in terms of a and p.

.... **(2 marks)**

Rearranging formulae

D **1** You can use this formula to work out the total cost, £C, for hiring a car for n days.
$C = 25n + 40$
Jerry pays £265 to hire a car.
For how many days did Jerry hire the car?

Guided

$C = 25n + 40$

Substitute the cost of £265 into the formula in place of C.

$\dots\dots\dots = 25n + 40$ $(-\ 40)$

$\dots\dots\dots = 25n$ $(\div\ 25)$

$n = \dots\dots\dots$

Jerry hired the car for .... days. **(2 marks)**

D **2** $P = 2a + 3b$
Work out the value of a when $P = 40$ and $b = 4$

$a =$ .... **(2 marks)**

D **3** $5a - 2b = m$
Work out the value of a when $m = 42$ and $b = -6$

$a =$ .... **(2 marks)**

C **4** Make t the subject of the formula $v = u + 6t$

Guided

$v = u + 6t$ $(-\ u)$

$v - \dots\dots\dots = 6t$ $(\div\ 6)$

$\frac{v - \dots\dots}{\dots\dots} = t$

$t = \frac{v - \dots\dots}{\dots\dots}$ **(2 marks)**

C **5** Make a the subject of the formula $2a - t^2 = c$

$a =$ .... **(2 marks)**

C **6** Make k the subject of the formula $p = 2(k - 3)$

Start by multiplying out the brackets.

$k =$ .... **(3 marks)**

Coordinates

F **1** (a) Write down the coordinates of

(i) point A

(........... , ...) **(1 mark)**

(ii) point B.

(........... , ...) **(1 mark)**

> Remember the first number is the distance across, the second number is the distance up or down.

(b) On the grid,

(i) plot the point (1, 0). Label the point C.

(ii) plot the point (–3, 1). Label the point D.

(2 marks)

D **2** (a) Write down the coordinates of

(i) point P

(........... , ...) **(1 mark)**

(ii) point Q.

(........... , ...) **(1 mark)**

M is the midpoint of PQ.

(b) Find the coordinates of M.

(........... , ...) **(2 marks)**

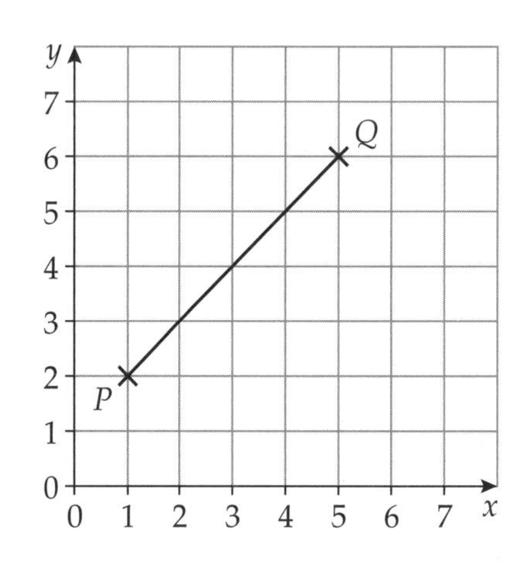

D **3** Work out the coordinates of the midpoint of the line joining (3, 4) and (7, 6).

Guided

Coordinates of midpoint

= (mean of x-coordinates, mean of y-coordinates)

$$= \left(\frac{3+7}{2}, \frac{\ldots\ldots + \ldots\ldots}{2}\right)$$

= (........... , ...)

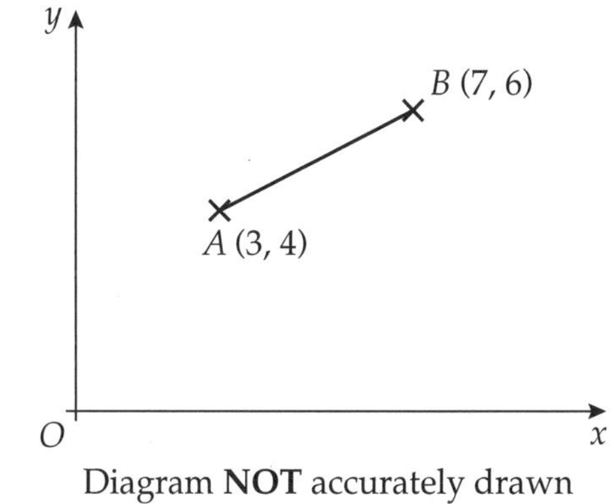

Diagram **NOT** accurately drawn

(2 marks)

C **4** Work out the coordinates of the midpoint of the line joining (0, 1) and (10, 5).

(........... , ...) **(2 marks)**

Straight-line graphs 1

D **1** The grid shows a straight line.
Work out the gradient of this straight line.

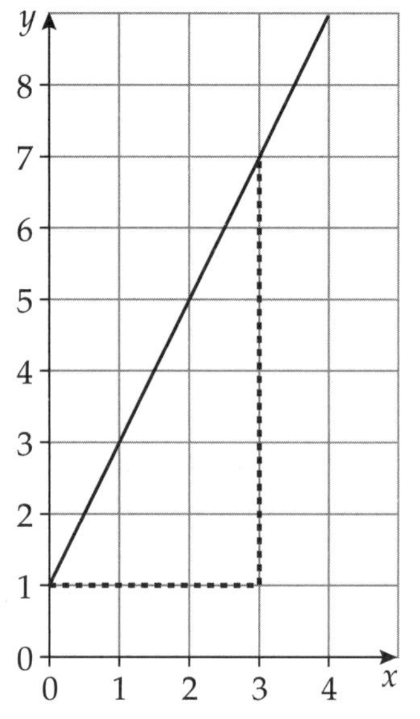

Guided Gradient $= \dfrac{\text{distance up}}{\text{distance across}} = \dfrac{\dots\dots\dots}{3}$

$= \dots\dots\dots$

Draw a triangle on the graph and use this to find the gradient.

(2 marks)

D **2** The grid shows a straight line.
Work out the gradient of this straight line.

The gradient of this graph is negative.

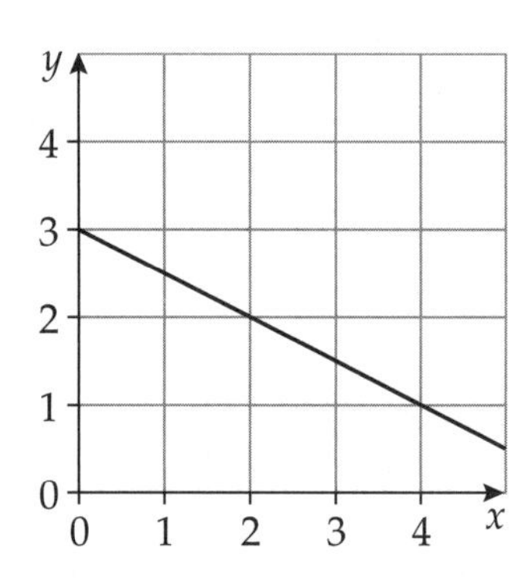

.... **(2 marks)**

C **3** The point A has coordinates (0, 20).
The point B has coordinates (8, 44).

(a) Work out the gradient of the line that passes through A and B.

.... **(2 marks)**

C is the point with coordinates (10, 10).

(b) Work out the gradient of the line that passes through B and C.

Look carefully at the scale on each axis.

.... **(2 marks)**

Straight-line graphs 2

D **1** (a) Complete the table of values for $y = 3x - 1$

> Substitute each value for x into the rule $y = 3x - 1$ to find the value of y.

x	−1	0	1	2	3
y		−1			8

Guided

$x = -1$: $y = (3 \times -1) - 1 =$ .... $- 1 =$ ....

$x = 1$: $y = (3 \times 1) - 1 =$ ....

(1 mark)

(b) On the grid, draw the graph of $y = 3x - 1$

> Plot your points on the graph and then draw a straight line through your points.

(2 marks)

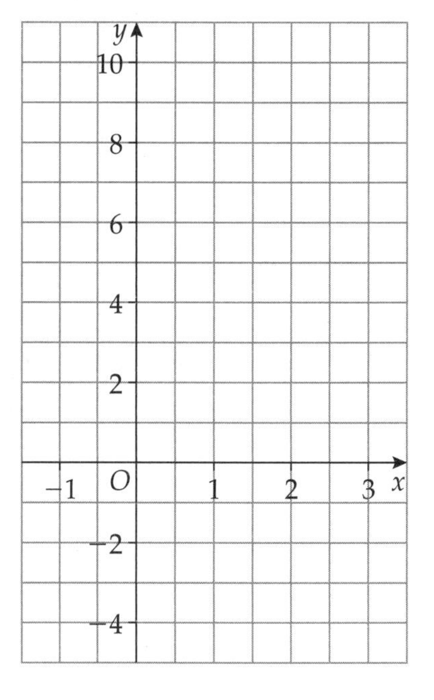

D **2** Draw the graph of $x + y = 6$ for values of x from −2 to 6

Guided

EXAM ALERT

x	−2	−1	0	1	2	3	4	5	6
y									

$x = -2$: $y = 6 - x = 6 - (-2) =$ ....

> Exam questions similar to this have proved especially tricky – be prepared! **ResultsPlus**

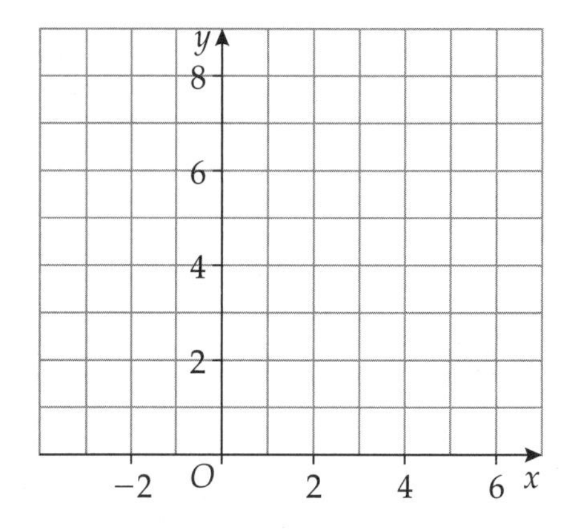

(3 marks)

3 (a) Draw the graph of $y = 2x + 5$ for values of x from −3 to 2

> First draw a table of values. The question tells you to use 'values of x from −3 to 2'. Next work out the values of y.

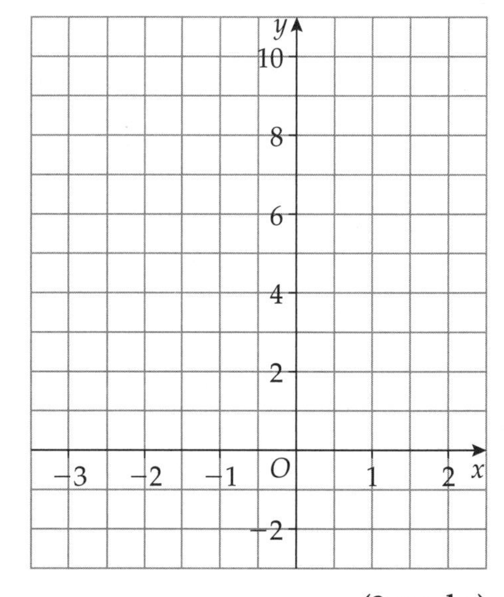

(3 marks)

(b) Work out the gradient of the line.

.... **(1 mark)**

Real-life graphs

D **1** You can use this graph to change between pounds and kilograms.

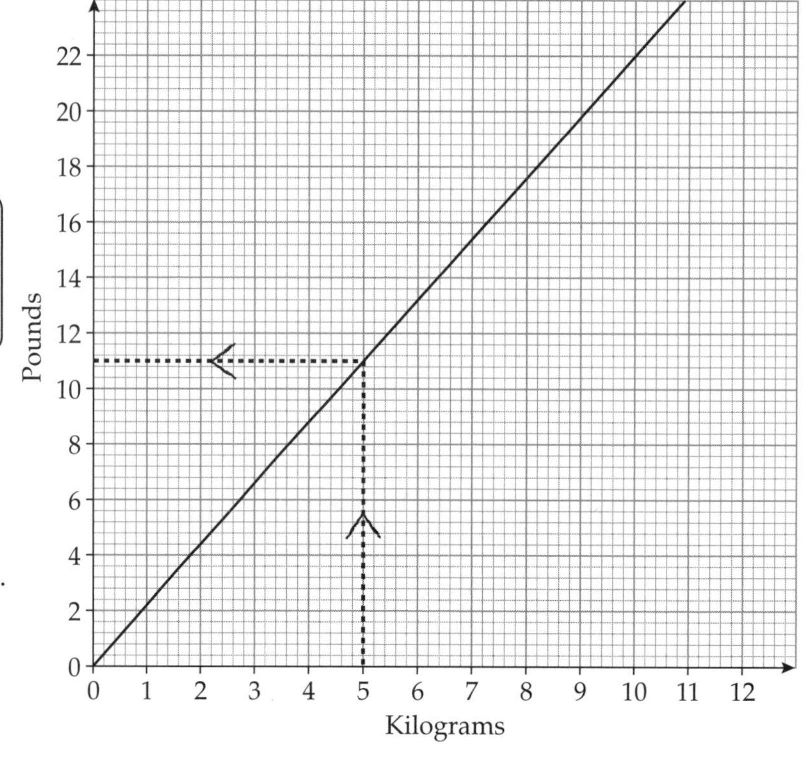

(a) Use the graph to change 5 kilograms into pounds.

Draw a vertical line from 5 kilograms to the line and then draw a horizontal line across to the pounds axis.

Guided

5 kilograms

= .... pounds

(1 mark)

(b) Use the graph to change 15 pounds into kilograms.

Guided

15 pounds

= .... kilograms

(1 mark)

Sue weighs 100 pounds.
Liz weighs 55 kilograms.

EXAM ALERT

(c) Who weighs more?
You must show all your working.

Exam questions similar to this have proved especially tricky – be prepared! ResultsPlus

Guided

10 pounds = ... kilograms

100 pounds = ... × 10 kilograms

= .... kilograms

.... weighs more than ....

100 pounds = 10 × 10 pounds

(3 marks)

D **2** You can use this graph to change between litres and gallons.

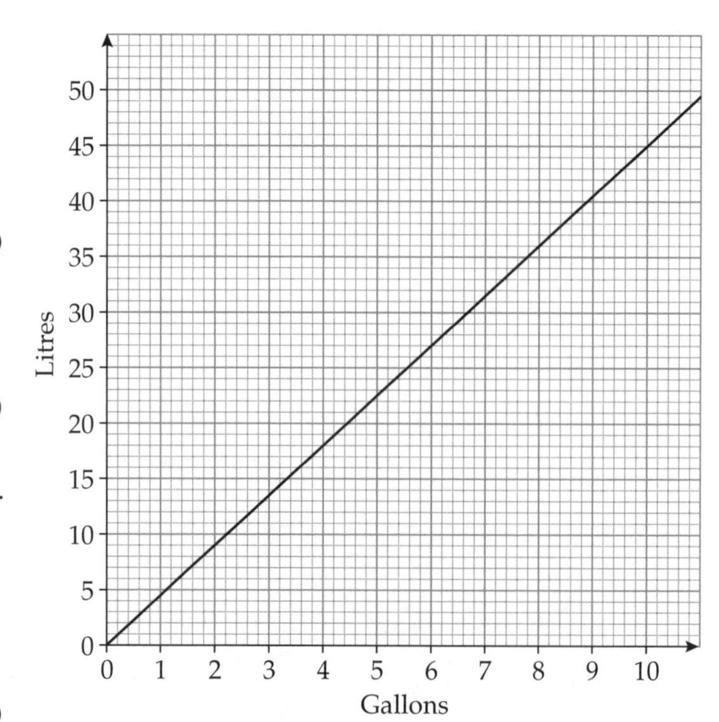

(a) Use the graph to change 8 gallons into litres.

.... litres **(1 mark)**

(b) Use the graph to change 27 litres into gallons.

.... gallons **(1 mark)**

(c) Change 120 litres into gallons.
You must show all your working.

.... gallons **(3 marks)**

Distance–time graphs

D **1** Jerry went on a cycle ride.
The travel graph shows Jerry's distance from home on this cycle ride.

(a) How far had Jerry cycled after 40 minutes?

> Draw a line up from 40 minutes then across to the distance axis.

Guided Distance after 40 minutes = ... km

(1 mark)

After 60 minutes, Jerry stopped for a rest.

(b) For how many minutes did he rest?

> Jerry is resting when the graph is horizontal.

Guided Jerry rested for .... minutes.

(1 mark)

Jerry left home at 09:30

(c) At what time did Jerry arrive back home?

Guided Journey time = ... minutes

= .... hours .... minutes

Jerry arrived back home at ....

(2 marks)

C **2** Here is part of a distance–time graph showing Kelly's journey from her house to the Sports Centre and back.

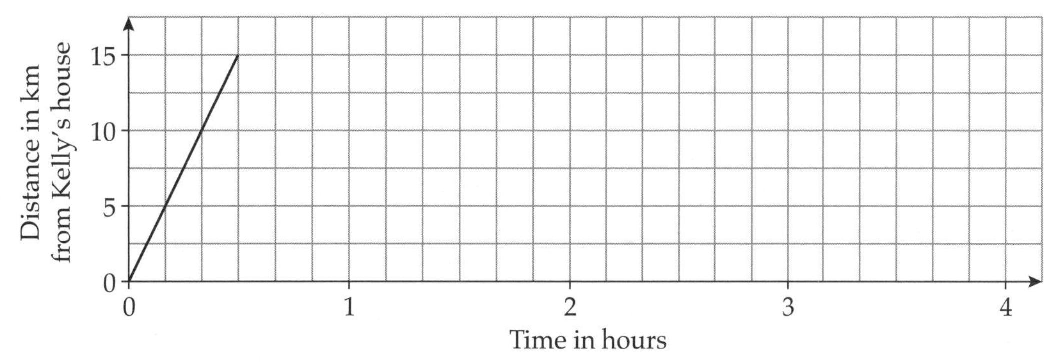

(a) Work out Kelly's speed for the first part of her journey.
Give your answer in km/h.

> Speed = $\frac{\text{distance}}{\text{time}}$
> Use time in hours.

.... km/h **(2 marks)**

(b) She spends 90 minutes at the Sports Centre and then travels home at a speed of 45 km/hour.
Complete the distance–time graph.

> Draw a horizontal line on the graph to show the time she spent at the Sports Centre.
> Use time = $\frac{\text{distance}}{\text{speed}}$ to work out the time it takes her to get home. Then draw a line on the graph to represent this part of her journey.

(2 marks)

Interpreting graphs

E **1** Sarah is in hospital.
The chart shows her temperature chart one day.

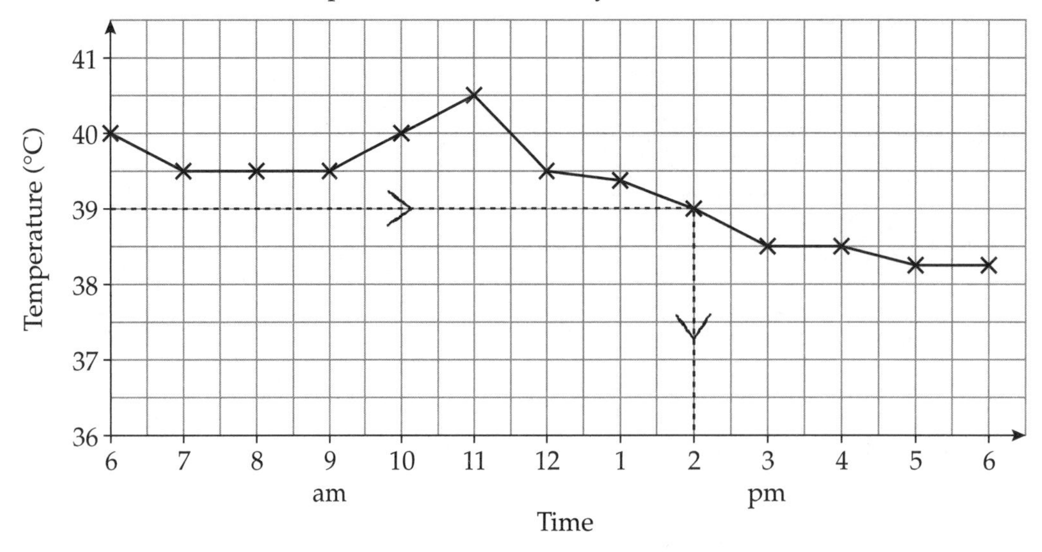

Guided (a) At what time was Sarah's temperature 39 °C?

........2 pm........ **(1 mark)**

(b) At what times was Sarah's temperature 39.5 °C?

.... **(1 mark)**

(c) Between which times did Sarah's temperature go **up** by 1 °C?

.... **(1 mark)**

(d) What can you say about Sarah's temperature between 11 am and 6 pm?

....

(1 mark)

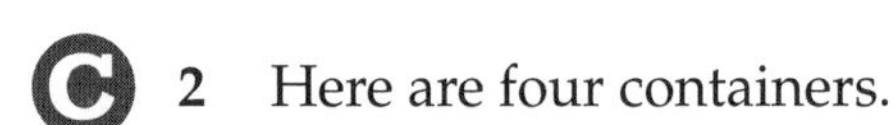

C **2** Here are four containers.
Water is poured into each container at a constant rate.

1

2

3

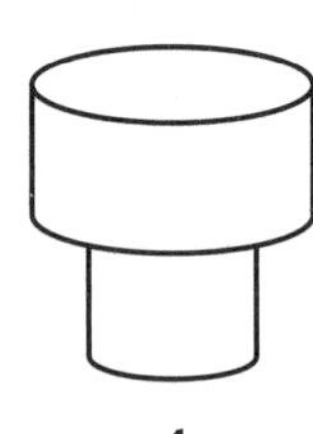

4

The narrower the width of container, the faster the water depth will increase.

Here are four graphs.
The graphs show how the depth of the water in each container changes with time.

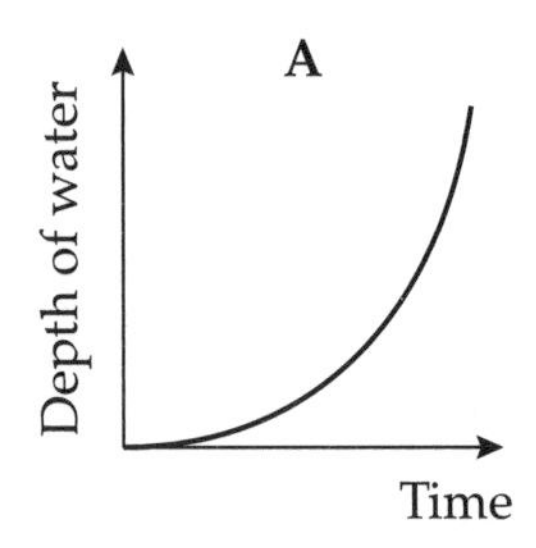

B

Depth of water

Time

C

Depth of water

Time

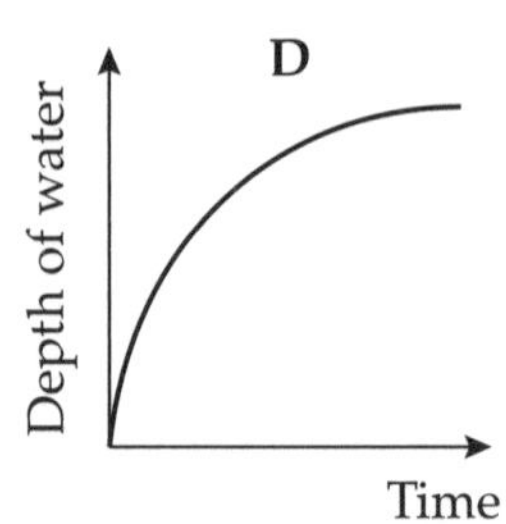

Match each graph with the correct container.

A and .... **B** and .... **C** and .... **D** and .... **(2 marks)**

Quadratic graphs

1 (a) Complete the table of values for $y = x^2 + 2x - 5$

x	−2	−1	0	1	2	3
y		−6	−5			10

$x = -2$ $y = (-2)^2 + 2 \times -2 - 5$

$= \ldots\ldots\ldots\ldots$

$x = 1$ $y = 1^2 + 2 \times 1 - 5$

$= \ldots\ldots\ldots\ldots$

$x = 2$ $y = 2^2 + 2 \times 2 - 5$

$= \ldots\ldots\ldots\ldots$

Write all the values in the table.

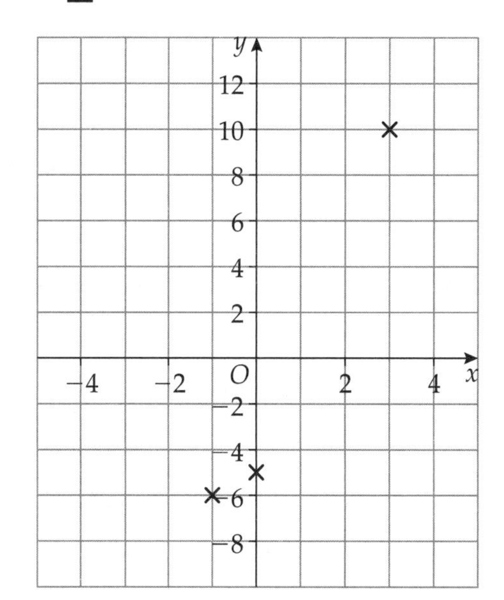

(2 marks)

(b) On the grid, draw the graph of $y = x^2 + 2x - 5$

Plot the points from the table and join with a smooth curve.

(2 marks)

2 (a) Complete the table of values for $y = x^2 - 3x - 1$

x	−2	−1	0	1	2	3
y	9		−1		−3	

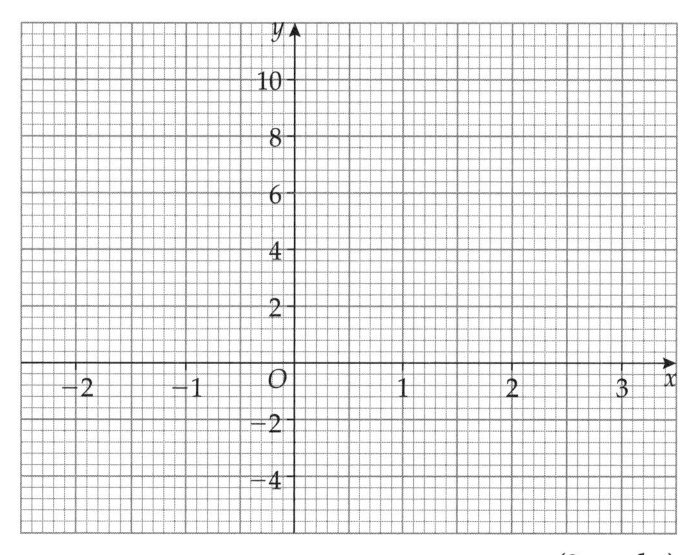

(2 marks)

(b) On the grid, draw the graph of $y = x^2 - 3x - 1$

(2 marks)

Using quadratic graphs

1 Here is the graph of $y = x^2 - 2x - 3$

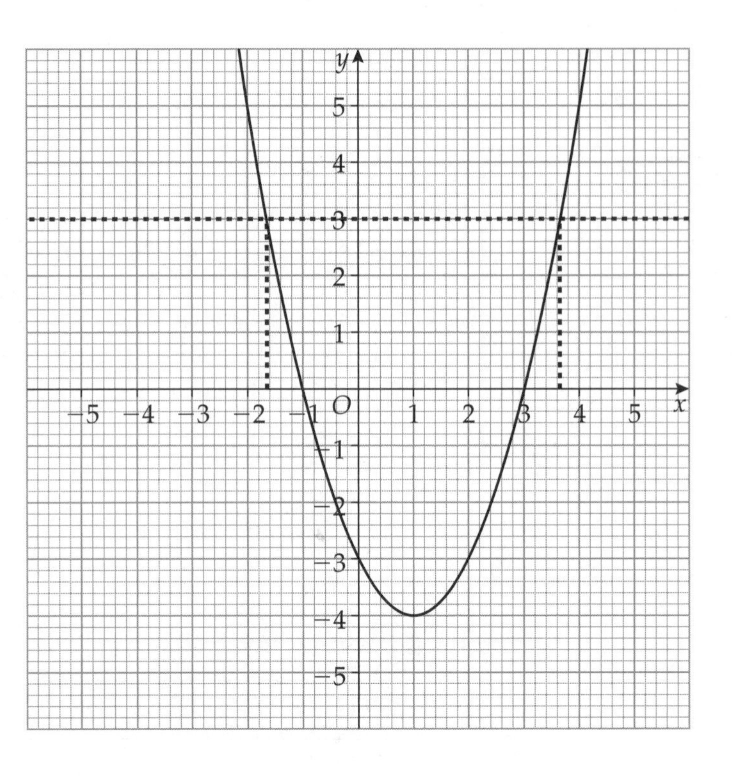

(a) Find the values of x when $y = 3$

> Draw the line $y = 3$ on the graph. Then read off the values of x.

Guided

$x =$ .... and $x =$ ....

(2 marks)

(b) Find the values of x when $y = -2$

> Draw the line $y = -2$ on the graph.

Guided

$x =$ .... and $x =$ ....

(2 marks)

(c) Use the graph to find the solutions of $x^2 - 2x - 3 = 0$

> Read off the values of x where the graph crosses the x-axis.

$x =$ .... and $x =$ ....

(2 marks)

2 Here is the graph of $y = 3 + x - x^2$

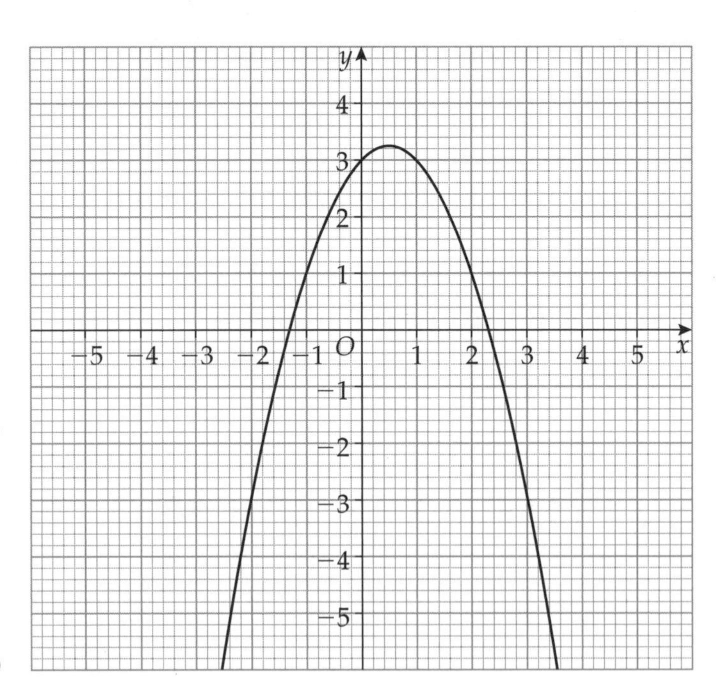

(a) Use the graph to find an estimate for the maximum value of y.

> Read off the value of x from the highest point on the graph.

$y =$ .... **(2 marks)**

(b) Find the values of x when $y = 1$

.... **(2 marks)**

(c) Use the graph to find the solutions of $3 + x - x^2 = 0$

.... **(2 marks)**

Problem-solving practice

E **1** Here are some patterns made from sticks.

Pattern number 1   Pattern number 2   Pattern number 3

(a) How many sticks are needed for Pattern number 6?

Draw a table showing the number of sticks in each pattern. Carry on until you get to Pattern number 6.

.... **(2 marks)**

(b) Harry says that he can make a pattern out of 49 sticks.
Is Harry correct? You must give a reason for your answer.

....

....

(2 marks)

E **2** You can use this graph to change between litres and pints.
Change 40 litres into pints.

Change 4 litres into pints and then use this result to change 40 litres into pints.

.... **(2 marks)**

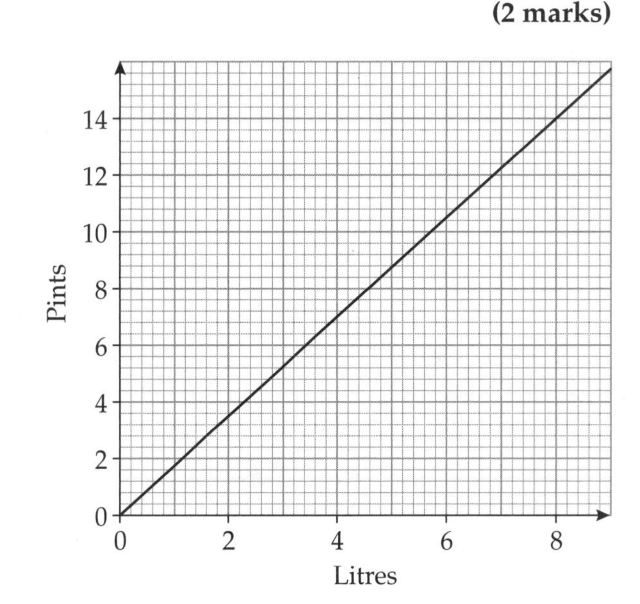

E **3** You can use this formula to work out the cost, in pounds, of having some curtains made.

Cost (in pounds) = length of curtains (in metres) × 35 + 60

Luke wants some curtains made. His curtains will have a length of 140 centimetres.

(a) Work out the cost.

140 cm = ... m

Cost = ... × .... + 60 = £........... **(3 marks)**

Beth had some curtains made. The cost was £147.50

(b) Work out the length of Beth's curtains.

To find the length, reverse the formula. Subtract 60 from the cost and then divide by 35

.... **(3 marks)**

Problem-solving practice

D *4 Here is a triangle.

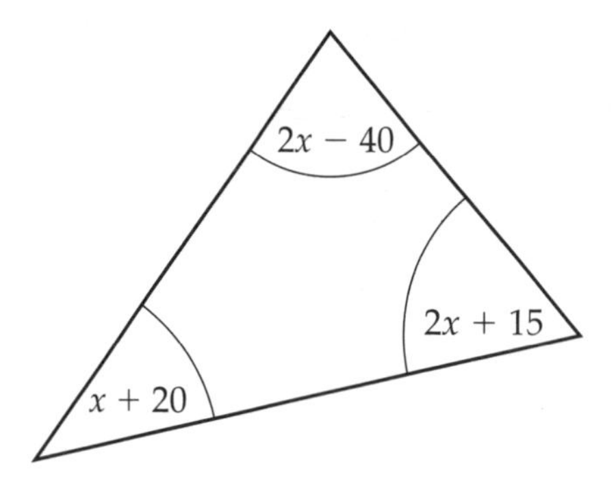

Diagram **NOT** accurately drawn

All angles are measured in degrees.
Work out the size of the largest angle.

$x + 20 + 2x + 15 + \dots\dots - \dots\dots = 180$

$\dots\dots x - \dots\dots = 180$

$\dots\dots x = 180 + \dots\dots$

$\dots\dots x = \dots\dots$

$x = \dots\dots$

Largest angle: $2x + 15 = 2 \times \dots\dots + 15$

$= \dots\dots°$

Use the fact that angles in a triangle add up to 180°

(4 marks)

D 5 On the grid, draw the graph of $y = 2 - 3x$ for values of x from -1 to 3

Start by drawing a table of values for x from -1 to 3
Use the equation $y = 2 - 3x$ to work out the corresponding values of y.
Plot your points and then join them with a straight line.

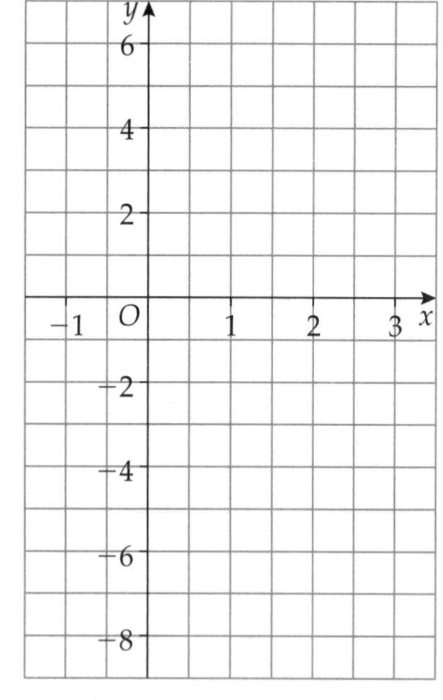

(3 marks)

C 6 Here is a triangle.
All the measurements are in centimetres.

x + 3
3x − 1
2x + 5

Diagram **NOT** accurately drawn

The triangle has a perimeter of 22 cm.
Work out the length of the shortest side of the triangle.

Write down an equation for the perimeter of the triangle.

.... **(4 marks)**

Measuring and drawing angles

1 (a) Measure the size of angle a.

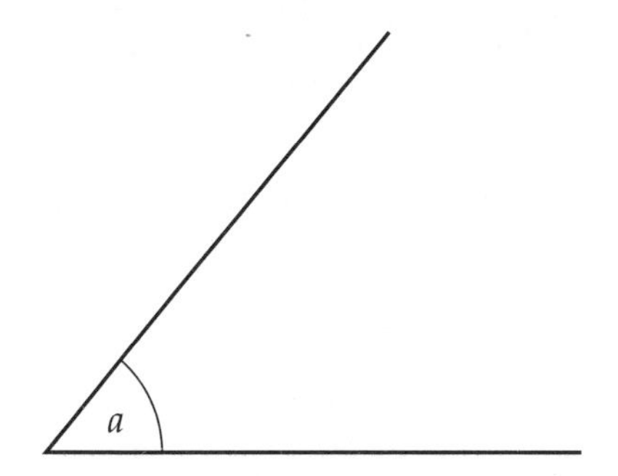

First estimate the size of the angle.

....................° **(1 mark)**

(b) Measure the size of the angle marked b.

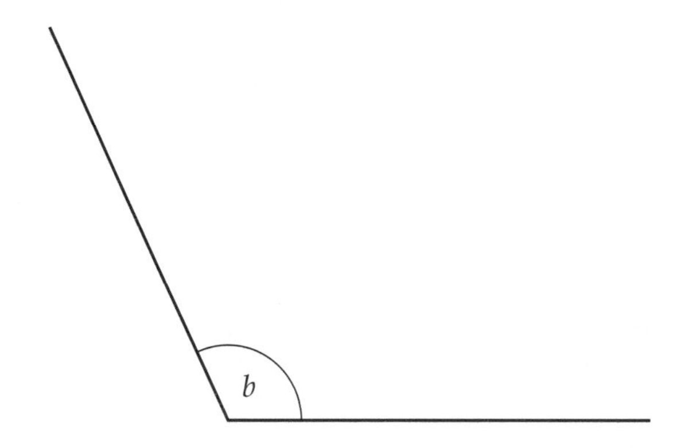

....................° **(1 mark)**

2 In the space below,

(a) accurately draw an angle of 68°

(b) accurately draw an angle of 130°.

(1 mark)

(1 mark)

Angles 1

G **1** (a) What type of angle is shown by the letter x?

> Angle x is bigger than a right angle but smaller than a straight line.

.... **(1 mark)**

(b) What type of angle is shown by the letter y?

.... **(1 mark)**

F **2** (a) Work out the size of the angle marked p.

Guided

$124 + p =$ ....

$p =$ .... $- 124$

$=$ ....°

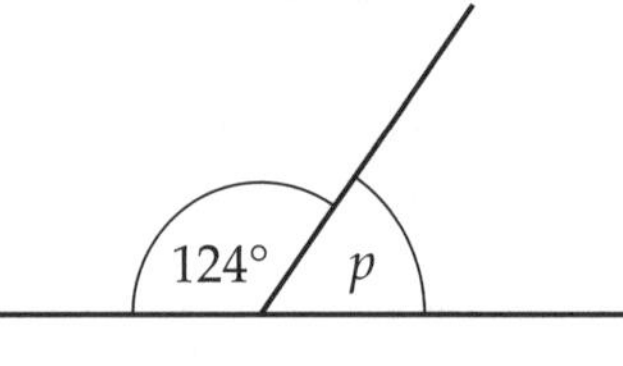

Diagram **NOT** accurately drawn

(1 mark)

(b) Give a reason for your answer.

Guided Angles on a straight line add up to .... **(1 mark)**

F **3** (a) Work out the size of the angle marked y.

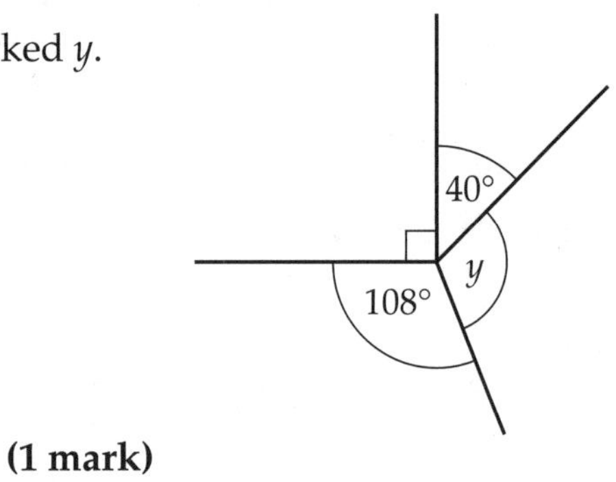

Diagram **NOT** accurately drawn

.....................° **(1 mark)**

(b) Give a reason for your answer.

....

(1 mark)

E **4** AB is a straight line.
This diagram is wrong. Explain why.

> Add up the two angles. Give an angle fact as part of your explanation.

Diagram **NOT** accurately drawn

(1 mark)

Angles 2

1 Work out the size of angle a.
Give a reason for your answer.

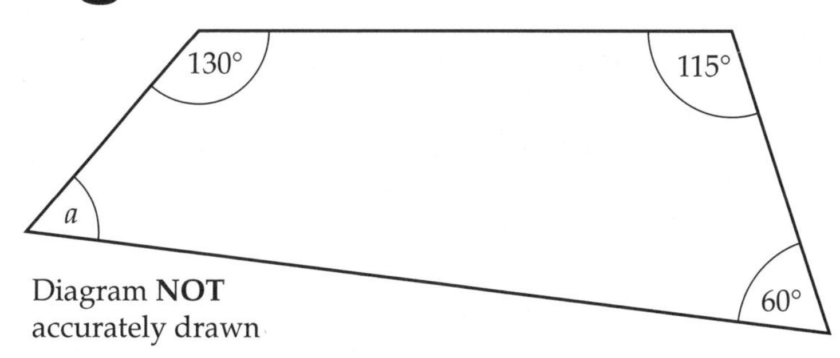

....................° **(3 marks)**

2 $ABCD$ is a straight line.
PQ is parallel to RS.

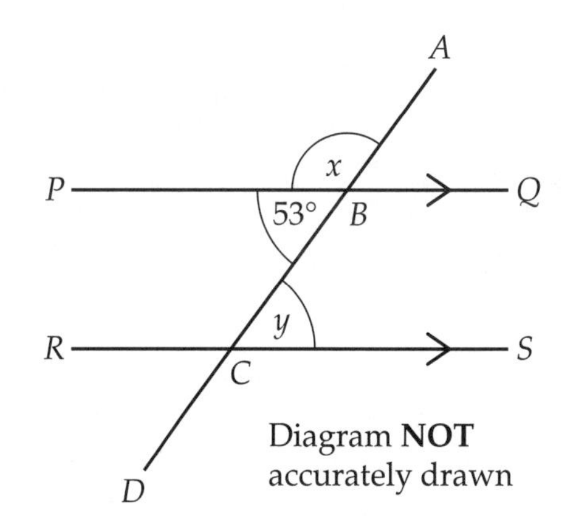

(a) Write down the size of the angle marked x.
Give a reason for your answer.

$x =$ ....°

Angles on a straight line add up to ....°

(2 marks)

(b) Work out the size of the angle marked y.
Give a reason for your answer.

$y =$ ....°

.... angles are equal.

(2 marks)

3 AB is parallel to CD.
EF is a straight line.
Write down the value of y.
Give a reason for your answer.

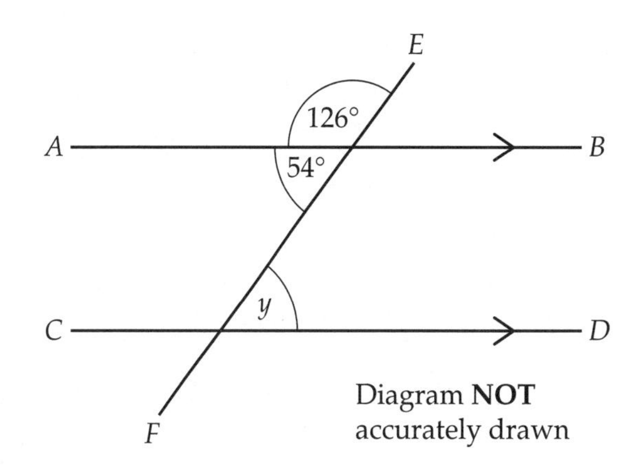

....................° **(2 marks)**

Solving angle problems

E **1** Triangle ABC is isosceles.
$AB = BC$
Angle $C = 47°$.
Work out the value of x.
Give reasons for your answer.

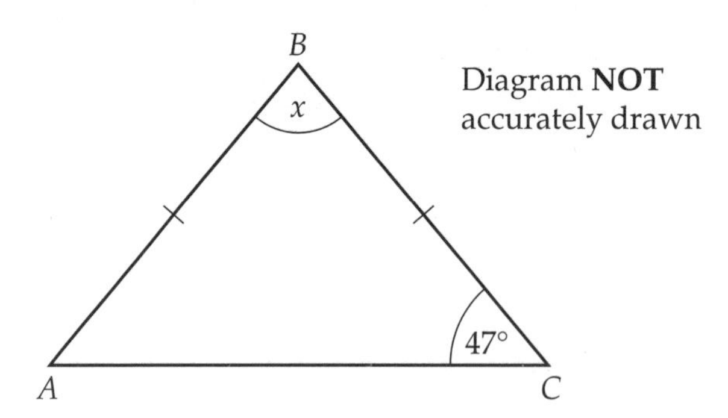

.....................° **(3 marks)**

D **2** PQ is parallel to RS.
OSQ and ORP are straight lines.

(a) Write down the value of x.
Give a reason for your answer.

$x =$ ....°

.... angles are equal.

(2 marks)

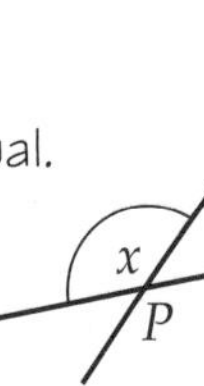

Q
85°
S
Diagram **NOT** accurately drawn
y
O
135°
R

(b) Work out the value of y.
Give reasons for your answer.

Guided

Angle $SRO = 180° - 135°$ — Angles on a straight line add up to ....°

$=$ ....°

Angle $RSO =$ ....° — .... angles are equal.

$y = 180° -$ ....$-$ .... — Angles in a triangle add up to ....°

$=$ ....°

(3 marks)

D **3** In the diagram, ABC is a straight line and $BD = CD$.
Work out the size of the angle marked y.
Give reasons for your answer.

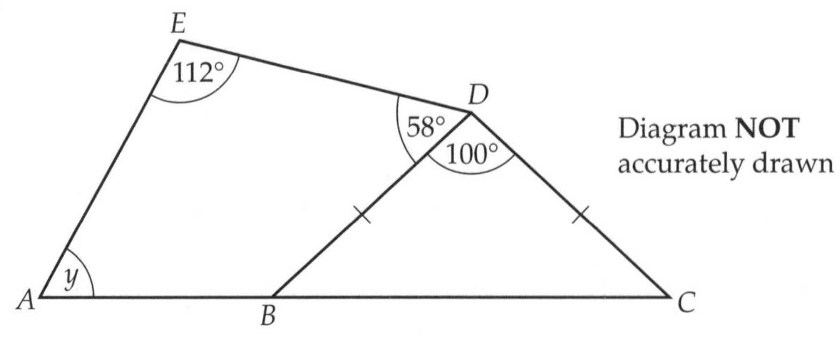

.....................° **(4 marks)**

Angles in polygons

D **1** Calculate the size of an exterior angle of a regular decagon.

Guided Sum of exterior angles of any polygon = ...°

Exterior angle of a regular decagon = $\frac{\text{...........}}{10}$ = ...° **(2 marks)**

D **2** The size of each exterior angle of a regular polygon is 45°.
Work out the number of sides of the polygon.

.... **(2 marks)**

D **3** A regular polygon has 15 sides.
Work out the size of an exterior angle of this regular polygon.

....................° **(2 marks)**

D **4** The size of each interior angle of a regular polygon is 160°.
Work out the number of sides of the polygon.

Guided Exterior angle of polygon = 180° − interior angle of polygon

= 180° − 160°

= ....°

Number of sides = $\frac{360}{\text{...........}}$ = ... **(2 marks)**

D **5** A regular polygon has 24 sides.
Work out the size of each interior angle of this regular polygon.

.... ° **(2 marks)**

C **6** The diagram shows a regular pentagon.
Work out the size of the angle marked x.
Give reasons for your working.

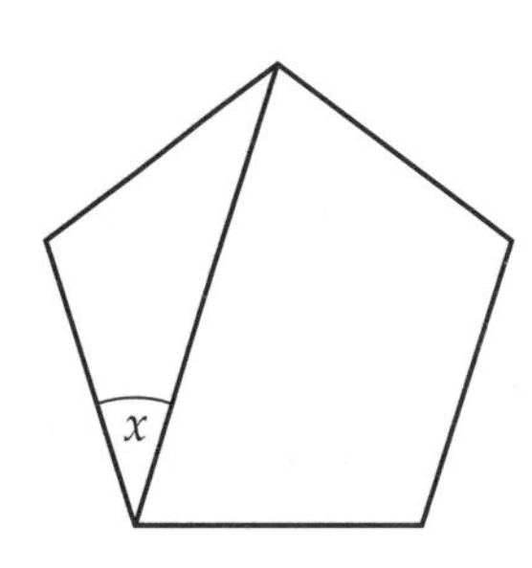

....................° **(4 marks)**

Measuring lines

G **1** (a) In the space below, draw a line 14 cm long.

(1 mark)

(b) Find the point that is halfway along the line you have drawn.
Mark it with a cross (×).

(1 mark)

G **2** Measure the length of the line.

Remember to include units with your answer.

.... **(2 marks)**

G **3** (a) Draw a line 7.5 cm long in the space below. Start from the point labelled A.

A
×

(1 mark)

(b) Mark with a cross (×), the point on your line which is 3 cm from the point A.

(1 mark)

(c) Mark the midpoint of the line PQ with a cross (×).

P ———————————————— Q

(1 mark)

F **4** The diagram shows a man standing next to a tree.
The man is of normal height.
The man and the tree are drawn to the same scale.

(a) Write down an estimate for the height, in metres, of the man.

.... m **(1 mark)**

(b) Work out an estimate for the height, in metres, of the tree.

Work out approximately how many times taller the tree is than the man.

.... m **(2 marks)**

Bearings

1 Guided

EXAM ALERT

Bearings are measured clockwise from North. Start by marking the angle you need to find. For part (a), measure the acute angle then subtract it from 360°.

Exam questions similar to this have proved especially tricky – be prepared! ResultsPlus

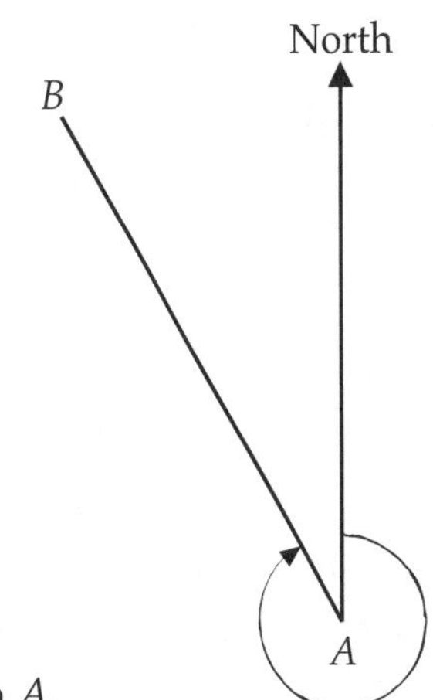

(a) Measure and write down the bearing of B from A.

.....................° **(1 mark)**

(b) On the diagram, draw a line on a bearing of 200° from B. **(1 mark)**

2 The bearing of Q from P is 140°.
What is the bearing of P from Q?

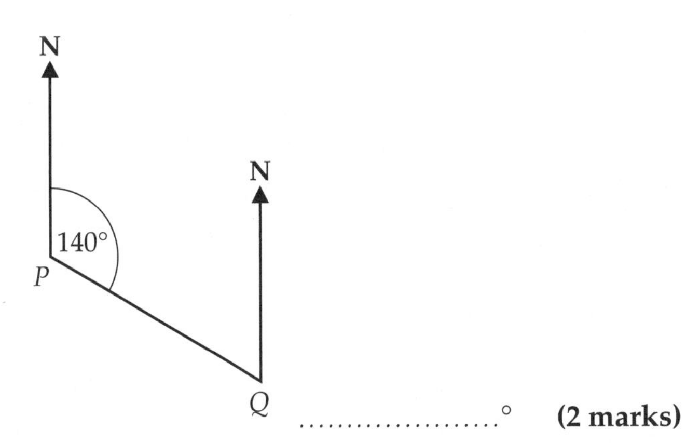

.....................° **(2 marks)**

3 The diagram shows the positions of two boats, P and Q.
The bearing of a boat R from boat P is 030°.
The bearing of boat R from boat Q is 300°.
Complete the diagram, making an accurate diagram to show the position of boat R.
Mark the position of boat R with a cross (×). Label it R.

Remember to measure bearings in a clockwise direction from North.

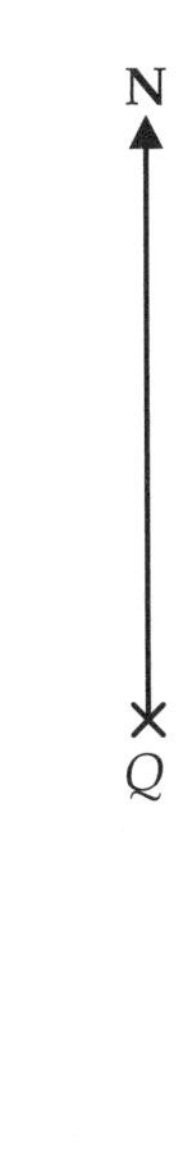

N

P

(3 marks)

Scale drawings and maps

D **1** A model of an aeroplane is made using a scale of 1 : 500

(a) The length of the model is 14.2 cm.
Work out the length of the aeroplane.
Give your answer in metres.

Guided

Length of aeroplane = 14.2 × ....

= .... cm = ... m **(2 marks)**

(b) The width of the aeroplane is 59.5 m.
Work out the width of the model.
Give your answer in centimetres.

Guided

Width of model = 59.5 ÷ ....

= .... m = ... cm **(2 marks)**

C **2** The scale diagram shows an island. A, B and C are three towns on the island.

Scale 1 : 400 000

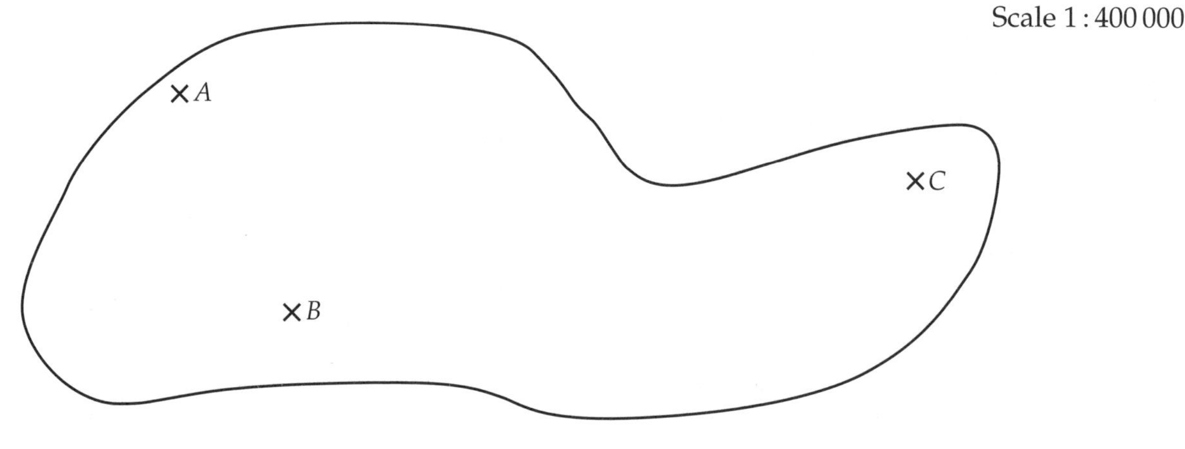

(a) Work out the actual distance from town A to town B. Give your answer in kilometres.

Guided

Distance on map from A to B = ... cm

Actual distance from A to B = ... × 400 000

= .... cm

= .... m

= .... km **(2 marks)**

(b) Work out the actual distance from town B to town C.

.... km **(2 marks)**

(c) Town D is 14 km away from town B.
Town D is on a bearing of 050° from town B.
Mark the position of town D on the map.

(4 marks)

Symmetry

1 A shaded shape has been drawn on a square grid.

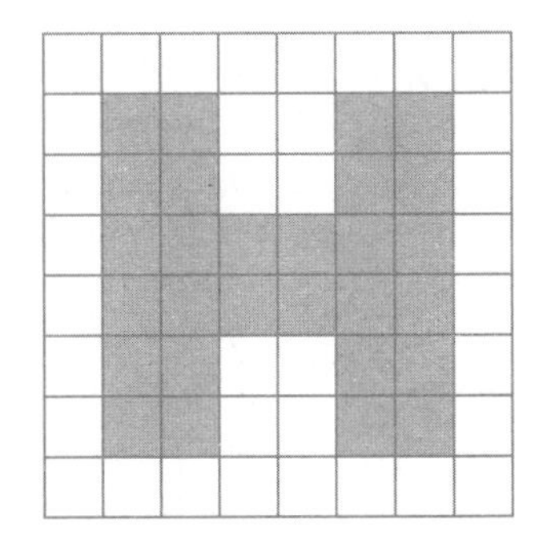

(a) Write down the order of rotational symmetry of the shaded shape.

Rotate the page and count how many times the shape fits over itself.

.... **(1 mark)**

The shaded shape has **two** lines of symmetry.

(b) Draw the lines of symmetry on the shaded shape. **(2 marks)**

2 Here are five shapes.

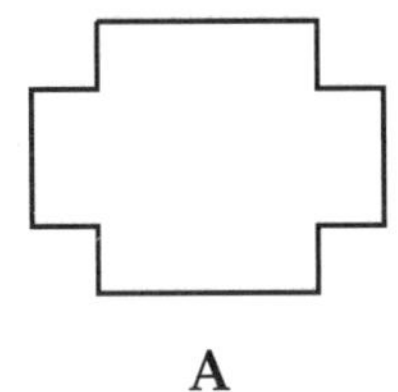

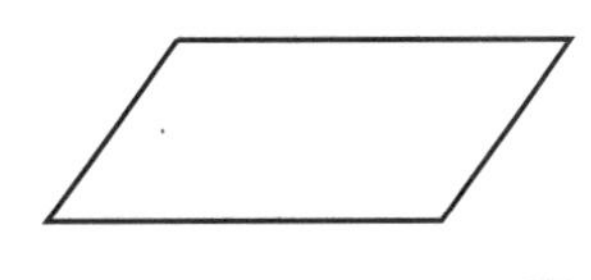

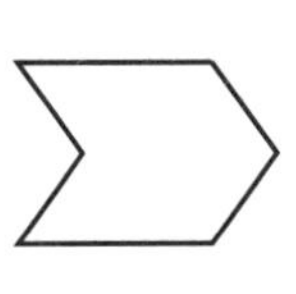

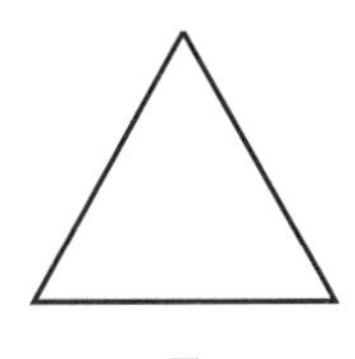

A **B** **C** **D** **E**

Two of these shapes have only **one** line of symmetry.

(a) Write down the letter of each of these **two** shapes.

.... and .... **(2 marks)**

Two of these shapes have rotational symmetry of order 2

(b) Write down the letter of each of these **two** shapes.

.... and .... **(2 marks)**

3 (a) On this diagram, shade **one** square so that the shape has exactly **one** line of symmetry.

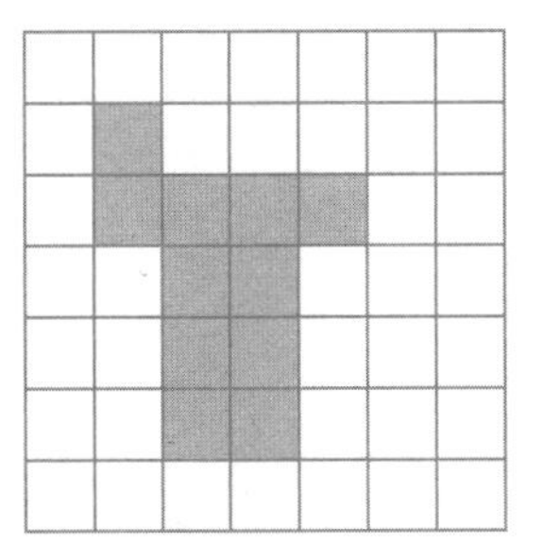

(1 mark)

(b) On this diagram, shade **one** square so that the shape has rotational symmetry of order 2

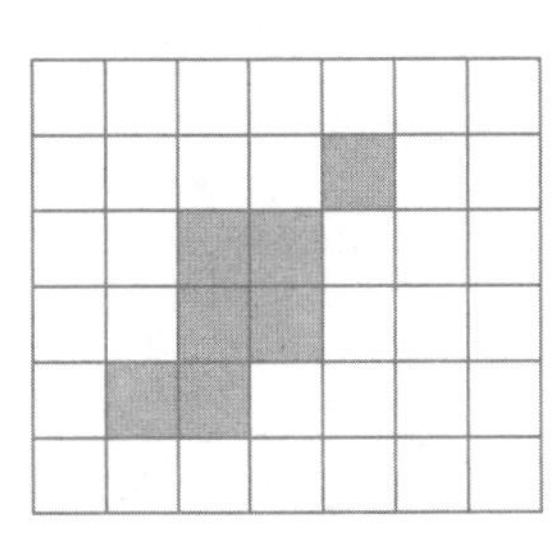

(1 mark)

2-D shapes

G 1 In each diagram the centre of the circle is marked with a cross (×).
Match each diagram to its label.

Circle and radius | Circle and chord | Circle and tangent | Circle and sector | Circle and diameter

Guided

(3 marks)

G 2 Write down the mathematical name for this quadrilateral.

.... (1 mark)

F 3 (a) Write down the mathematical name for each of these quadrilaterals.

(i)

This quadrilateral has only one pair of parallel lines.

.... (1 mark)

(ii)

This quadrilateral has two pairs of equal sides which are **not** parallel.

.... (1 mark)

(b) In the space below, draw a parallelogram.

A parallelogram has two pairs of equal parallel sides.

(1 mark)

F 4 Complete these sentences.

(a) A square has .... equal sides and .... equal angles. (1 mark)

(b) Each angle in a rectangle is ....°. (1 mark)

(c) A rectangle has .... lines of symmetry. (1 mark)

F 5 Write down the names of **three** quadrilaterals that have diagonals which intersect at 90°.

....

(2 marks)

Congruent shapes

G 1 Here are eight triangles.

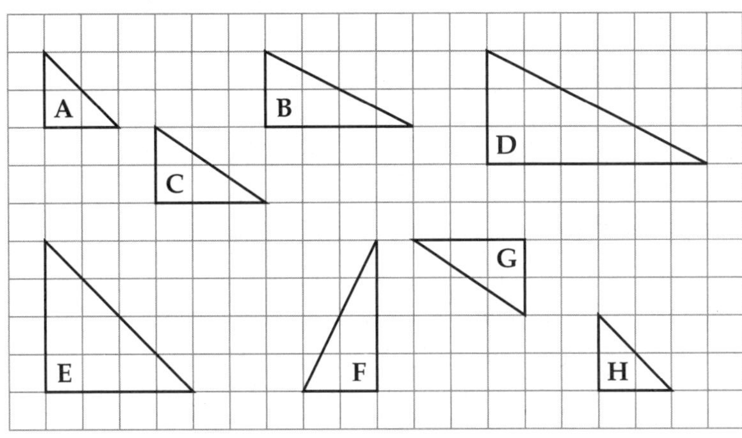

The triangle that is congruent to **A** will be exactly the same size and shape as triangle **A**.

(a) Write down the letter of a triangle that is congruent to triangle **A**.

.... **(1 mark)**

(b) Write down the letter of a triangle that is congruent to triangle **C**.

.... **(1 mark)**

(c) Write down the letter of a triangle that is congruent to triangle **F**.

.... **(1 mark)**

F 2 Here are two shapes.
Mary says, 'These two shapes are congruent.'
Is Mary right?
Give a reason for your answer.

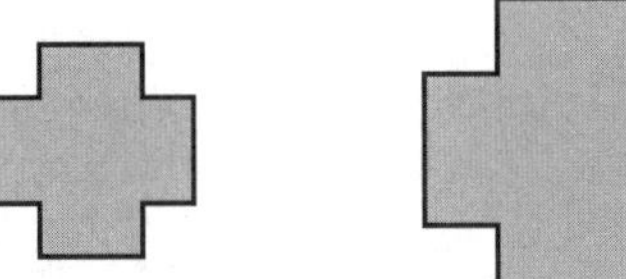

....

....

(1 mark)

E 3 On the grid, show how the shaded shape will tessellate.
You must draw at least six more shapes.

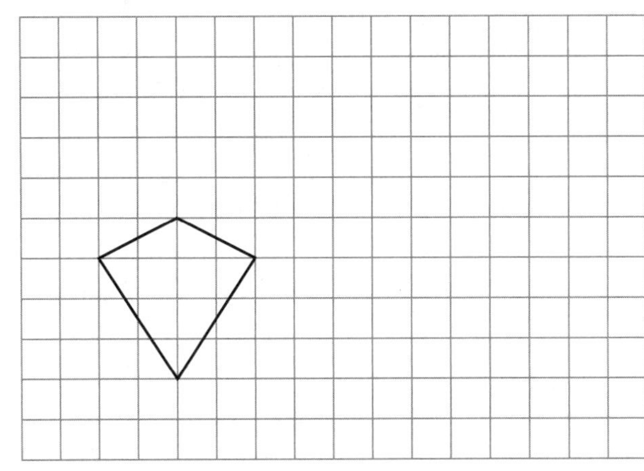

Draw at least six more shapes to cover the grid. There must not be any gaps between your shapes.

(2 marks)

Reading scales

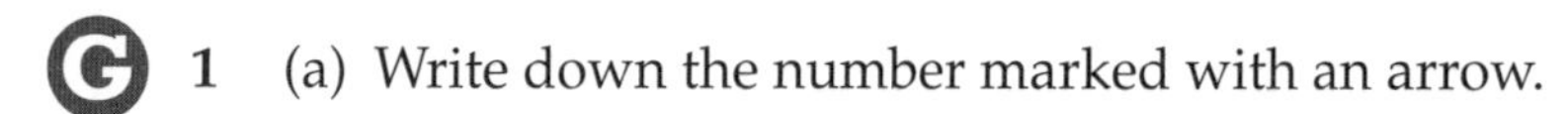

G **1** (a) Write down the number marked with an arrow.

.... **(1 mark)**

(b) Write down the number marked with an arrow.

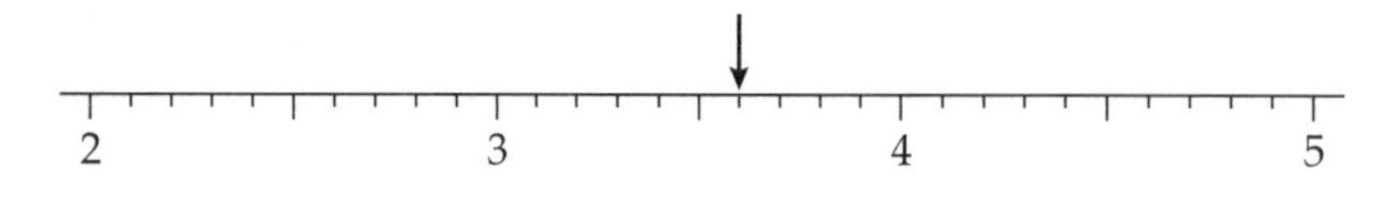

.... **(1 mark)**

(c) Find the number 420 on the number line. Mark it with an arrow (↑).

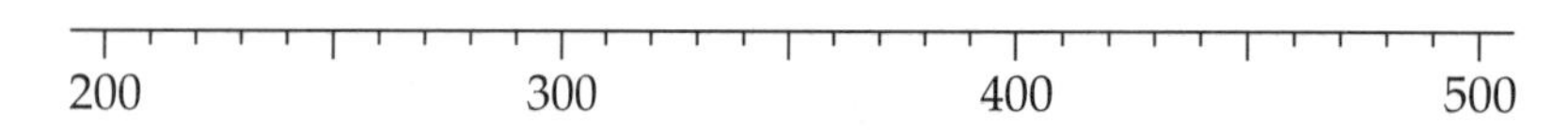

(1 mark)

(d) Find the number 3.8 on the number line. Mark it with an arrow (↑).

(1 mark)

F **2** Write down the reading on each of these scales.

(a)

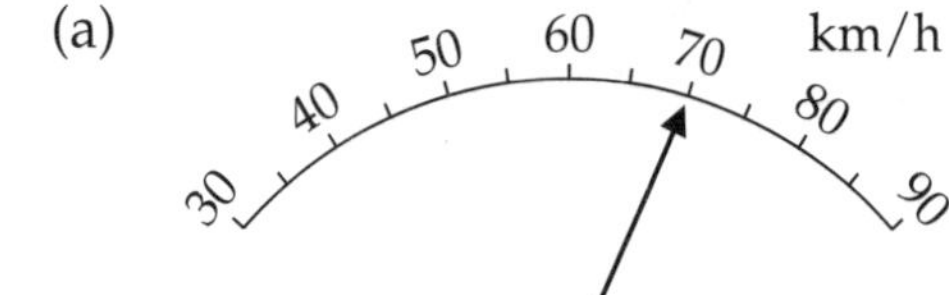

.... km/h

(b)

.... m*l*

(c)

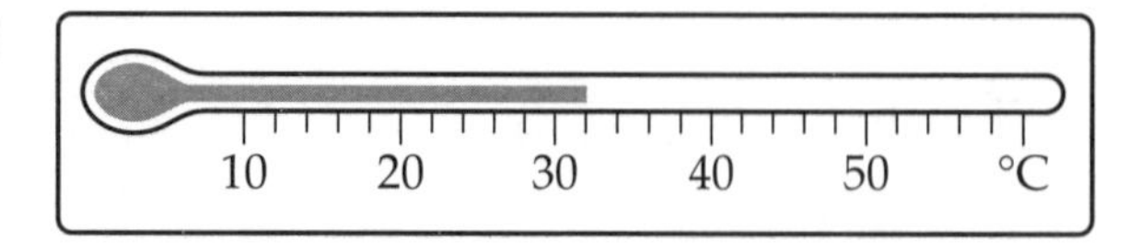

.... °C

(d)

.... kg

(4 marks)

Time and timetables

F **1** (a) Write 6.10 pm using 24-hour time.

.... **(1 mark)**

(b) A train left London at 13:30. The train arrived in Newcastle at 16:10
How long did the journey take?

Guided

13:30 to 14:00 → .... minutes

14:00 to 16:00 → .... hours

16:00 to 16:10 → .... minutes

Total journey time is .... hours and .... minutes **(1 mark)**

F **2** The table shows what Julie plans to do before arriving at work.

Activity	Time (min)
Drive from home to pool	10
Swim	45
Shower and change	20
Drive from pool to work	25

She has to arrive at work at 09:00
What is the latest time she can leave home?

.... **(3 marks)**

F **3** Here is part of a railway timetable.

Manchester	05:15	06:06	06:45	07:05	07:15	07:45
Stockport	05:26	06:16	06:55	07:15	07:25	07:55
Macclesfield	05:39	06:29	07:08		07:38	08:08
Stoke-on-Trent	05:54	06:45	07:24		07:54	08:24
Stafford	06:12		07:41		08:11	
London Euston	08:07	08:26	09:06	09:11	09:50	10:08

A train leaves Manchester at 07:15

(a) (i) At what time should this train get to London Euston?

.... **(1 mark)**

(ii) How long should this train take to travel between Manchester and London Euston?

.... **(1 mark)**

John lives in Macclesfield. He has to go to a meeting in London.
He needs to arrive in London Euston **before** 09:30

(b) What is the time of the latest train that he can catch from Macclesfield? .... **(1 mark)**

Kenny catches the 07:25 train from Stockport.
The train arrives in London Euston 40 minutes late.

(c) What time does the train arrive in London Euston?

.... **(2 marks)**

Metric units

G **1** (a) Change 4.5 m to centimetres.

1 m = 100 cm

.... cm **(1 mark)**

(b) Change 8900 g to kilograms.

1 kg = 1000 g

.... kg **(1 mark)**

(c) Change 450 mm into centimetres.

.... cm **(1 mark)**

(d) Change 4.08 km into metres.

.... m **(1 mark)**

E **2** Kim is making cakes.
She wants to make as many cakes as possible.
She needs 250 g of flour to make each cake.
Kim only has 1.5 kg of flour.
How many cakes can Kim make?

.... cakes **(3 marks)**

D **3** The weight of 3000 sweets is 1.2 kg.
Work out the weight of 1 sweet.
Give your answer in grams.

Guided

1 kg = ... grams

Weight of 3000 sweets = 1.2 × ....

= .... grams

Weight of 1 sweet = ... ÷ ....

= .... grams **(2 marks)**

D **4** A water tank has a volume of 40 litres.
The water tank is empty.
The tank is filled at a rate of 80 m*l* per second.
How long will the tank take to fill?
Give your answer in minutes and seconds.

.... minutes .... seconds **(2 marks)**

D **5** Daniel buys 1 kg of cheese for £13.50 at a farm shop.
He buys 200 g of the same cheese for £2.60 at a supermarket.
Is the cheese better value for money at the farm shop or at the supermarket?
You must show your working.

.... **(3 marks)**

Measures

E **1** John weighed 8.5 pounds when he was born.
Mark weighed 4 kg when he was born.
At birth, which baby weighed more?

> You should know the fact that 1 kg = 2.2 pounds.

Guided Mark weighed 4 kg = 4 × ....

= .... pounds

Therefore, ... weighed more at birth. **(3 marks)**

E **2** Change 64 kilometres into miles.

> You should know the fact that 8 km = 5 miles.

Guided 64 kilometres = $\frac{64}{\ldots\ldots}$ × 5 miles

= .... miles **(2 marks)**

D **3** A car travels at a speed of 35 miles per hour.
Change 35 miles per hour to kilometres per hour.

.... km/h **(2 marks)**

C **4** Kalim puts 10 gallons of petrol into his car's petrol tank.
Petrol costs £1.30 per litre.
How much did it cost Kalim to fill up his petrol tank?

> You should know the fact that 1 gallon = 4.5 litres.

£.................... **(3 marks)**

C **5** Sue drives at an average speed of 80 km/hour.
How long will it take her to drive 100 miles?

.... hours **(3 marks)**

C **6** Lottie buys a 1.5 litre bottle of drink.
She pours all the drink into glasses.
Each glass has a capacity of $\frac{1}{2}$ pint.
How many glasses will she be able to fill completely?

.... **(3 marks)**

Speed

D **1** The distance from London to Munich is 900 km.
A flight from London to Munich takes 3 hours.
Work out the average speed of the aeroplane.

Guided Speed $= \frac{\text{distance}}{\text{time}}$

$= \frac{\ldots\ldots\ldots\ldots}{\ldots\ldots\ldots\ldots}$

$=$ ………… km/h **(2 marks)**

C **2** Lottie drives at an average speed of 80 km/h.
Her journey took $2\frac{1}{2}$ hours.
Work out the length of Lottie's journey.

…………………. km **(2 marks)**

C **3** A train travelled 180 km in $1\frac{1}{2}$ hours.
Work out the average speed of the train.

$180 \div 1\frac{1}{2} = 360 \div 3$ as both sides have been multiplied by 2

…………………. km/h **(2 marks)**

C **4** An aeroplane travelled at an average speed of 600 km/h for 2 hours 15 minutes.
How far did the aeroplane travel?

Use the fact that 15 minutes is $\frac{1}{4}$ of an hour.

…………………. km **(2 marks)**

C **5** Jamil walked at an average speed of 6 km/h for 40 minutes.
How far did he walk?

Write the time as a fraction of an hour and simplify the fraction.

Guided Time $= \frac{40}{60}$ hour

$= \frac{\ldots\ldots\ldots\ldots}{\ldots\ldots\ldots\ldots}$ hour

Distance = speed × time

$= 6 \times \frac{\ldots\ldots\ldots\ldots}{\ldots\ldots\ldots\ldots}$

$=$ …………km **(2 marks)**

C **6** Jim leaves home at 14:30
He cycles 39 km to a friend's house at an average speed of 18 km/h.
At what time does he arrive at his friend's house?

…………………. **(2 marks)**

Perimeter and area

G 1 The shaded shape has been drawn on a grid of centimetre squares.

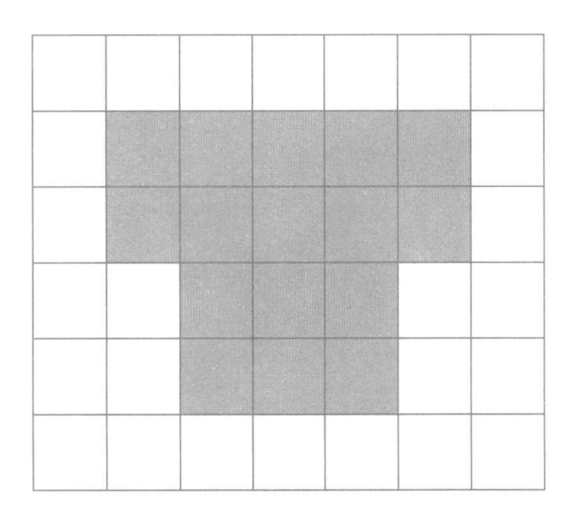

(a) Find the area of the shaded shape. Give units with your answer.

Count the number of shaded squares.

.... **(2 marks)**

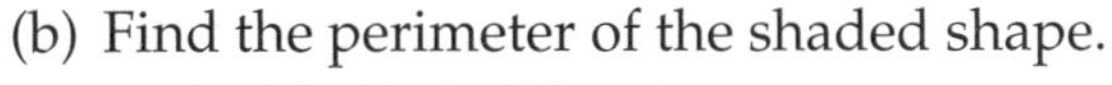

(b) Find the perimeter of the shaded shape.

Find the distance around the edge of the shaded shape.

.... cm **(1 mark)**

G 2 The shaded shape has been drawn on a grid of centimetre squares.

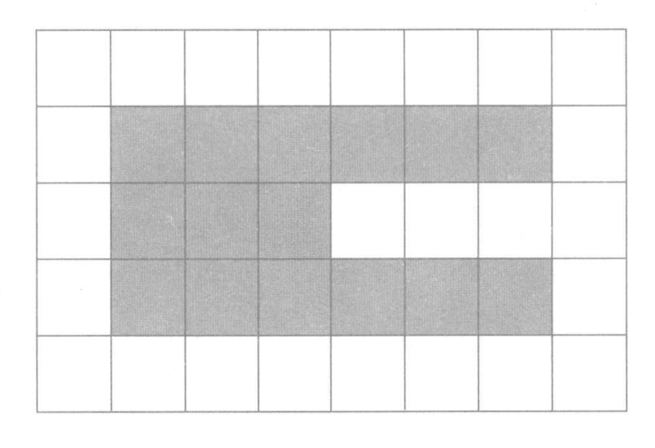

(a) Find the area of the shaded shape.

.... cm^2 **(2 marks)**

(b) Find the perimeter of the shaded shape. Give units with your answer.

.... **(1 mark)**

F 3 Work out the perimeter of this trapezium.

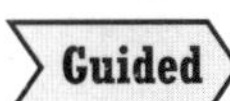

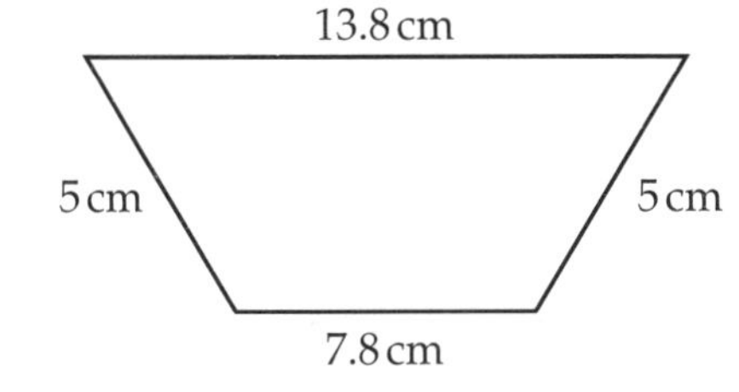

Perimeter = 13.8 + 5 + 7.8 + 5

= .... cm **(2 marks)**

F 4 Work out the perimeter of this rectangle.

Remember to include all four sides of the rectangle in your calculation.

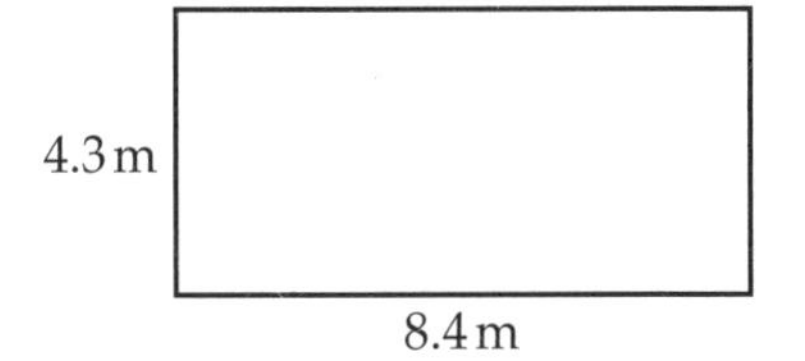

.... m **(2 marks)**

E 5 Work out the perimeter of this parallelogram.

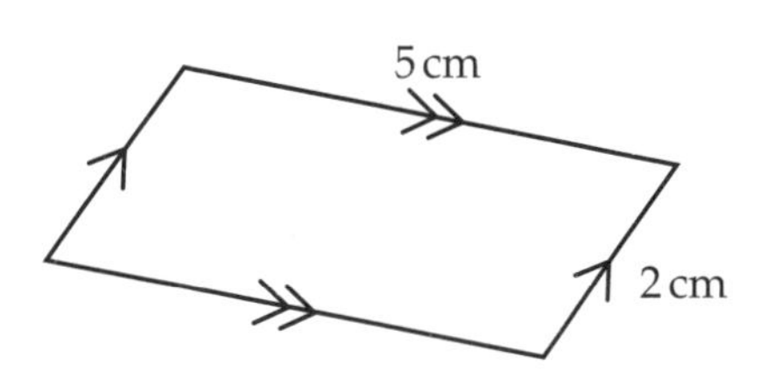

.... cm **(2 marks)**

Using area formulae

E **1** Work out the area of this rectangle.

5 cm

8 cm

.... cm^2 **(2 marks)**

E **2** The diagram shows a path.
The path is in the shape of a rectangle.
Work out the area of the path.
Give your answer in square centimetres.

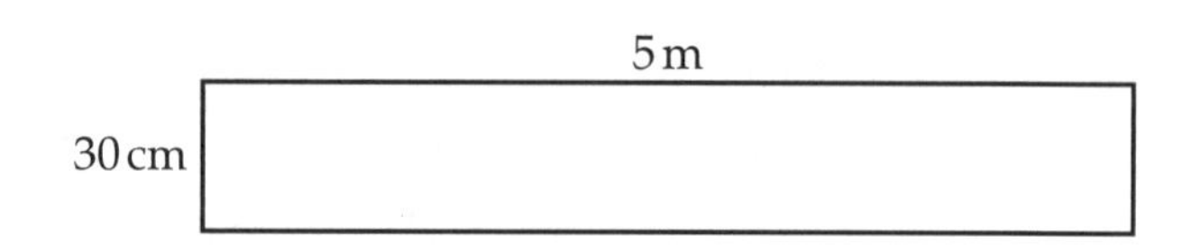

Guided

5 m = 5 × .... cm

= .... cm

Area of path = ... × ....

= .... cm^2 **(3 marks)**

D **3** Work out the area of the triangle.

Remember to divide by 2: $\frac{1}{2} \times 10 \times 4$ is the same as $10 \times 4 \div 2$

4 cm

10 cm

Guided

Area of triangle = $\frac{1}{2}bh$

= $\frac{1}{2}$ × .... × ....

= .... cm^2 **(2 marks)**

D **4** Work out the area of the triangle.
Give your answer in square centimetres.

Change 1.2 m into cm first.

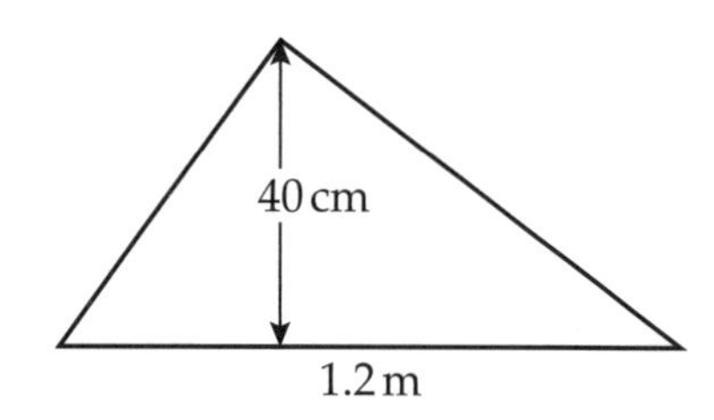

.... **(3 marks)**

C **5** The diagram shows a trapezium of height 8 cm.
Work out the area of this trapezium.

The formula for the area of a trapezium is given on the formula sheet: $A = \frac{1}{2}(a + b)h$

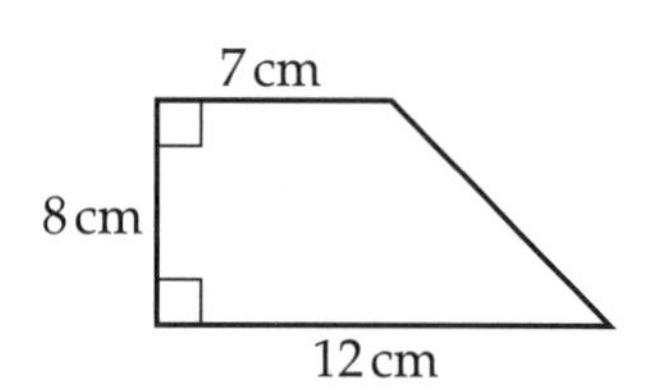

.... cm^2 **(3 marks)**

Solving area problems

D **1** Work out the area of this shape.

10 cm
A
6 cm
B
3 cm
4 cm

Divide the shape up into two rectangles.

Guided

Area of rectangle = length × width

Area of rectangle A = ... × ....

= .... cm^2

Area of rectangle B = ... × ....

= .... cm^2

Area of shape = ... + ...

= .... cm^2

(3 marks)

C **2** The diagram shows the plan of a room in a Community Centre.
Mr Foster wants to cover the floor with varnish.
One tin of varnish will cover 20 m^2.
How many tins of varnish will he need?

12 m
10 m
4 m
20 m

Divide the shape up into a triangle and a rectangle.

Guided

Area of rectangle = length × width

= .... × ....

= .... m^2

Height of triangle = 10 − 4 = ... m

Base of triangle = 20 − 12 = ... m

Area of triangle $= \frac{\text{base} \times \text{height}}{2}$

$= \frac{\text{...........} \times \text{...........}}{2}$ = ... m^2

Area of shape = ... + ...

= .... m^2

Number of tins of varnish = ... ÷ 20

= .... tins

The answer must be a whole number of tins.

(5 marks)

3 The diagram shows the plan of a field.
The farmer wants to sell the field.
He wants to get at least £4 per square metre.
Mr Hobbs offers him £20 000 for the field.
Will the farmer accept his offer?

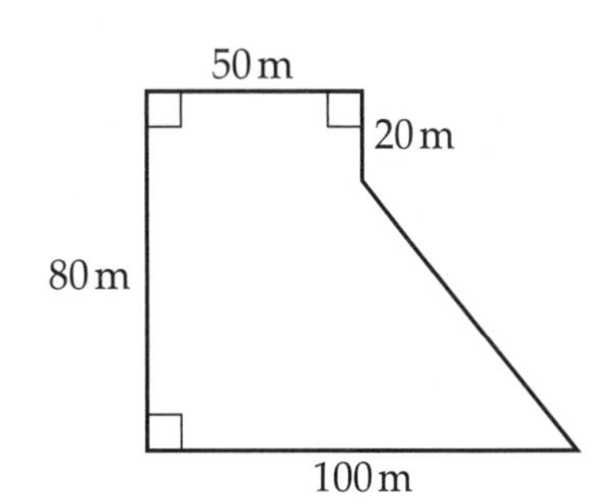

....

(5 marks)

Circles

D **1** A circle has a radius of 6.5 cm.
Work out the circumference of this circle.
Give your answer correct to 1 decimal place.

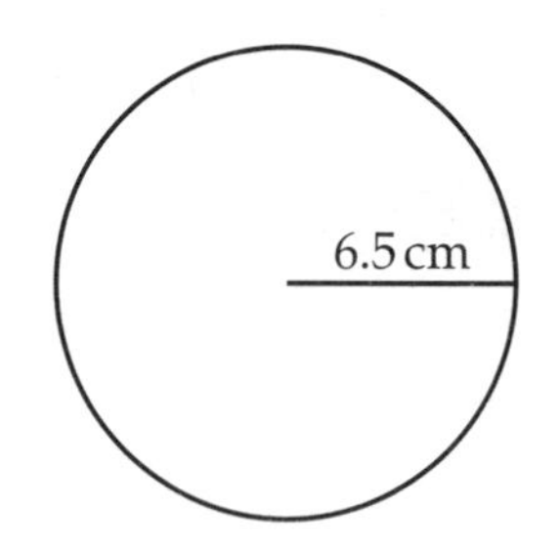

Guided

radius = 6.5 cm

diameter = ... × 6.5

= .... cm

Circumference of a circle = π × diameter

= π × ....

= ....

= .... cm to 1 d.p.

Write down at least six figures from your calculator.

(2 marks)

D **2** A circle has a radius of 8 cm.
Work out the circumference of this circle.
Give your answer correct to 3 significant figures.

.... cm **(2 marks)**

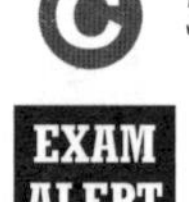

C **3** The diagram shows a semicircle.
The diameter of the semicircle is 18 cm.
Work out the perimeter of the semicircle.
Give your answer correct to 2 decimal places.

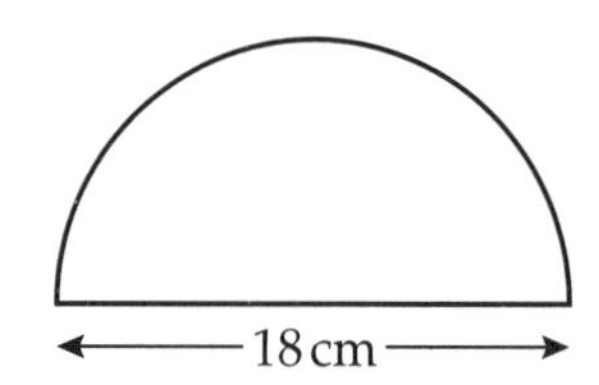

Guided

Perimeter of semicircle = $\frac{1}{2}$ × perimeter of circle + diameter

= $\frac{1}{2}$ × π × d + d

= $\frac{1}{2}$ × π × .... + ...

= .... + ...

= ....

= .... cm to 2 d.p.

Exam questions similar to this have proved especially tricky – be prepared! ResultsPlus

(3 marks)

4 A quarter circle has a radius of 6 cm.
Work out the perimeter of the quarter circle.
Give your answer correct to 1 decimal place.

The perimeter is made up of the arc and two radii.

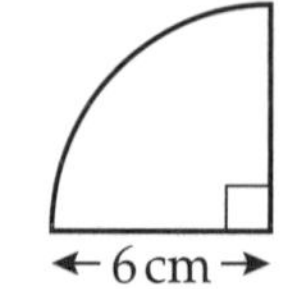

.... cm **(3 marks)**

Area of a circle

D **1** A circle has a radius of 7 cm.
Work out the area of this circle.
Give your answer correct to 1 decimal place.

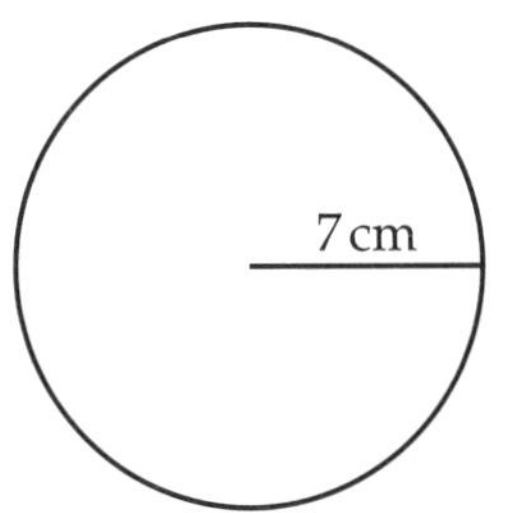

Guided

Area of circle $= \pi \times r^2$

$= \pi \times$ ....^{2}

$= \pi \times$ ....

$=$ ....

$=$ .... cm^2 to 1 d.p.

Work out r^2 first before multiplying by π.

Write down at least six figures from your calculator.

(2 marks)

D **2** A circle has a diameter of 18.4 cm.
Work out the area of the circle.
Give your answer correct to 3 significant figures.

First work out the radius of the circle.

.... cm^2 **(2 marks)**

C **3** The diagram shows a semicircle.
The diameter of the semicircle is 24 m.
Work out the area of the semicircle.
Give your answer correct to 3 significant figures.

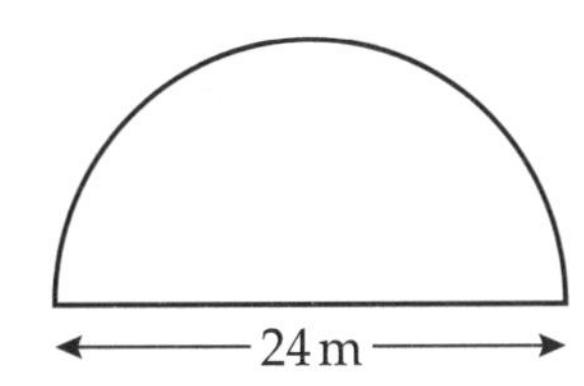

Guided

diameter $= 24$ m

radius $=$ .... m

Area of semicircle $= \frac{1}{2} \times$ area of circle

$= \frac{1}{2} \times \pi \times r^2$

$= \frac{1}{2} \times \pi \times$ ....^{2}

$=$ ....

$=$ ....m^2 to 3 s.f.

(3 marks)

C **4** Craig cuts a quarter circle from a piece of card.
The card is a square of sides 16 cm.
The radius of the quarter circle is 16 cm.
Work out the area of the leftover card.
Give your answer correct to 3 significant figures.

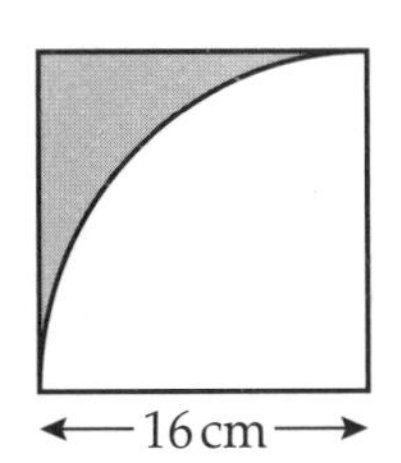

.... cm^2 **(4 marks)**

3-D shapes

G **1** Write down the name of each of these three 3-D shapes.

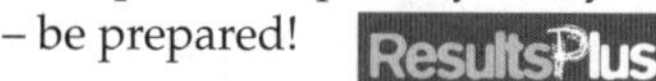

(a) 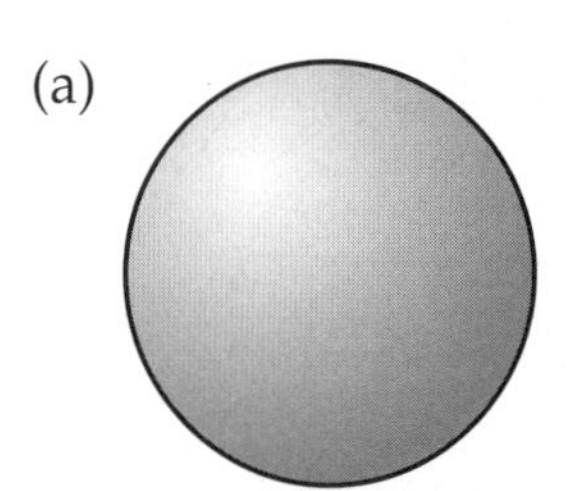

(b) 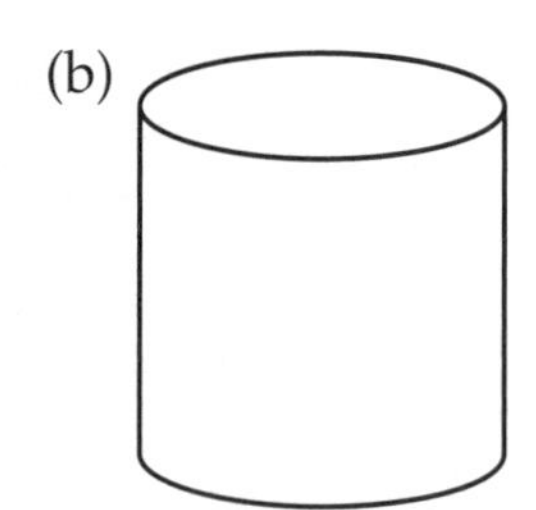

(c) 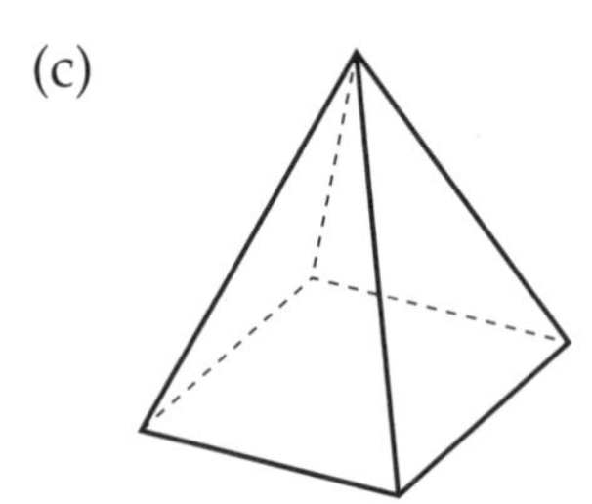

Guided

....Cylinder...... ....

(3 marks)

2 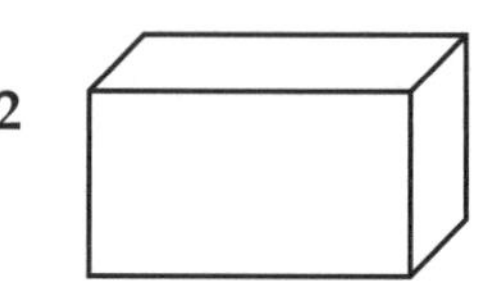

The diagram shows a cuboid.
Write down the number of

(a) faces (b) edges (c) vertices

....

(3 marks)

3 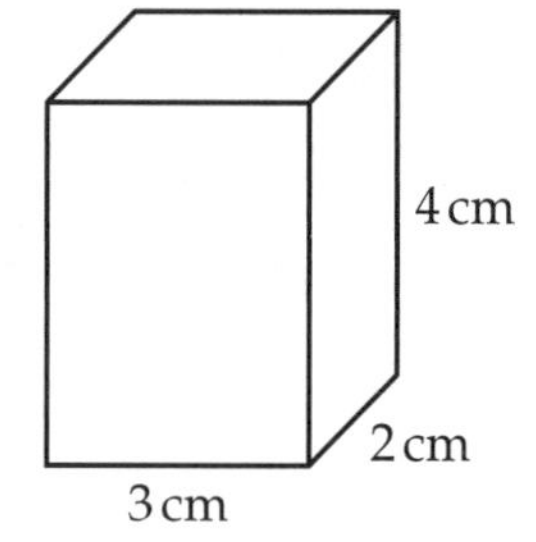

Here is a cuboid.

On the isometric paper, make an accurate drawing of the cuboid.

(3 marks)

4 The diagram shows a prism.
Write down the number of

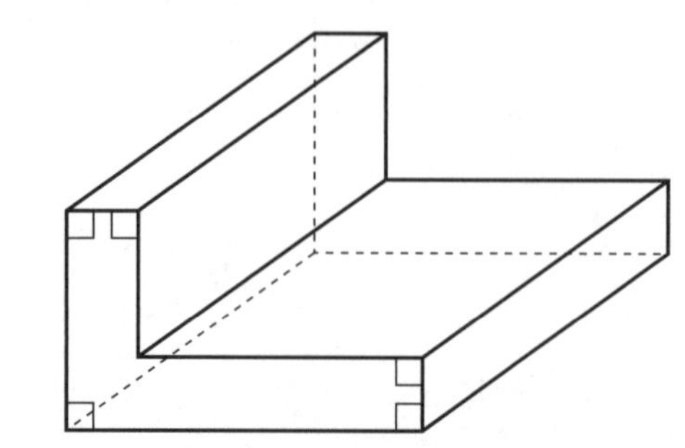

(a) faces ....

(b) edges ....

(c) vertices .... **(3 marks)**

Plan and elevation

1 The diagrams show some solid shapes and their nets.
Draw an arrow from each of the solid shapes to its net.

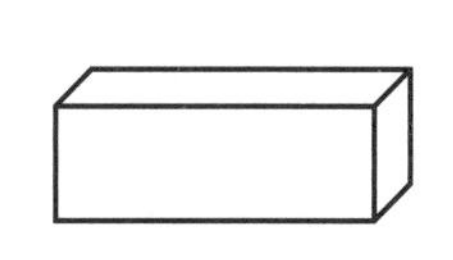

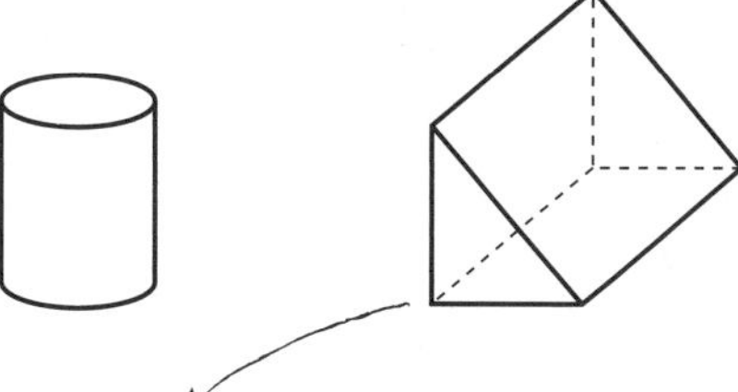

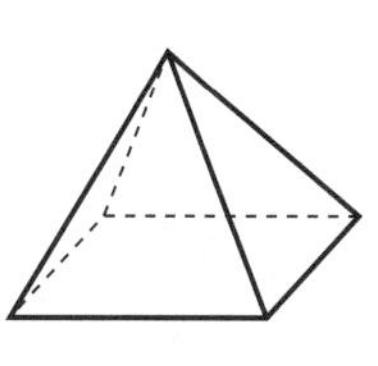

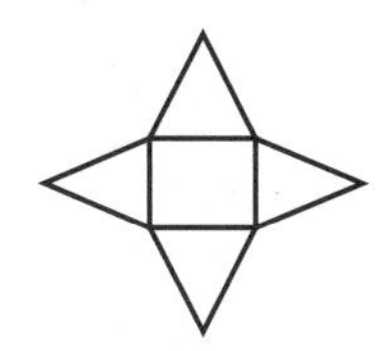

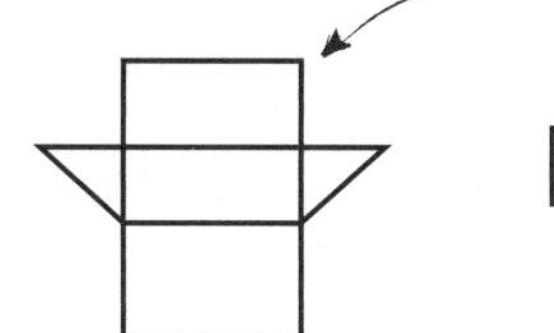

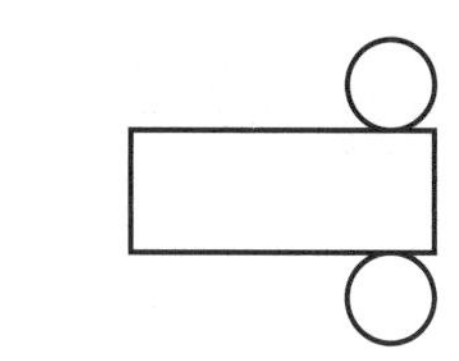

(3 marks)

2 The diagram shows a sketch of a solid object.
The solid object is made from 6 centimetre cubes.

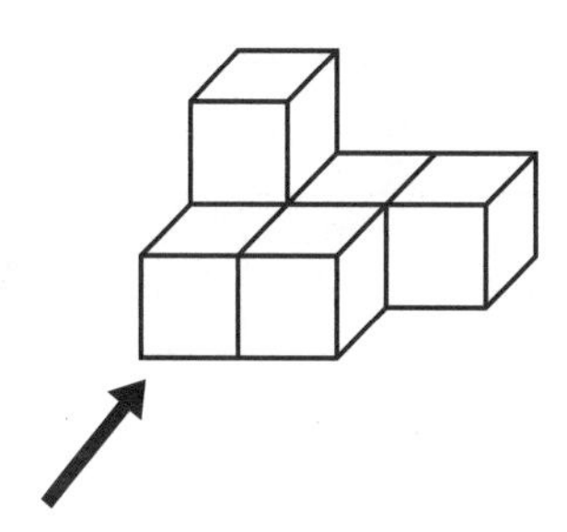

(a) Draw a sketch of the elevation of the solid object in the direction marked by the arrow.

(2 marks)

(b) Draw a sketch of the plan of the solid object.

(2 marks)

3 Here are the plan and front elevation of a solid shape.

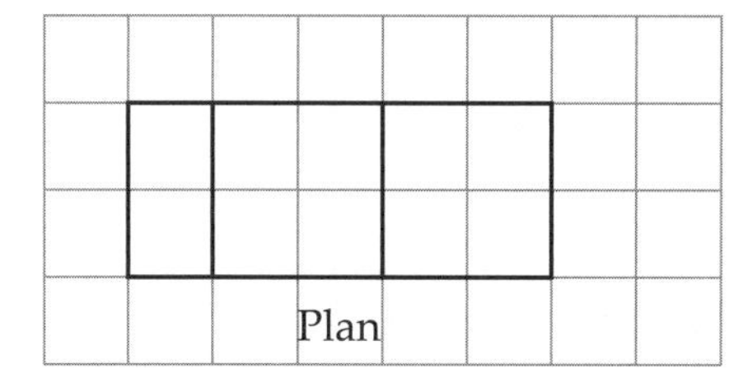

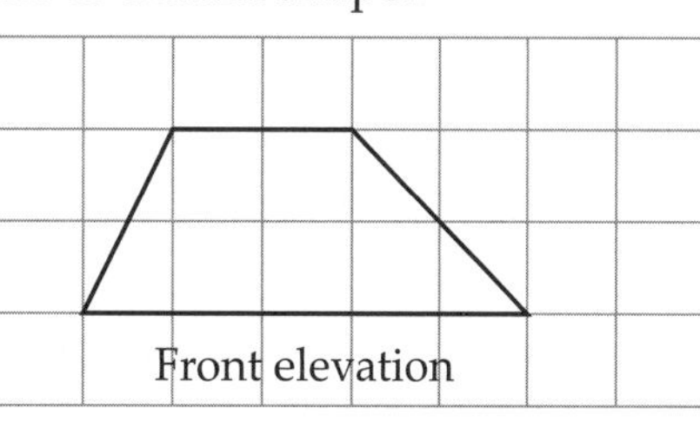

(a) On the grid below, draw the side elevation of the solid shape. **(2 marks)**

(b) Draw a 3-D sketch of the solid shape.

(2 marks)

Volume

G **1** This solid prism is made from centimetre cubes.
Find the volume of the prism.

Guided

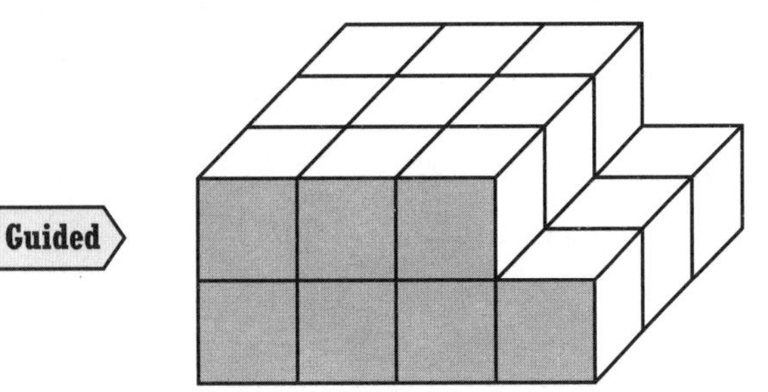

Count the cubes on the front face carefully.

Number of cubes on front of prism = ...

Volume = ... × 3

= .... cm^3

The prism is made up of 3 'faces'. Each 'face' has the same number of cubes.

(2 marks)

E **2**

Guided

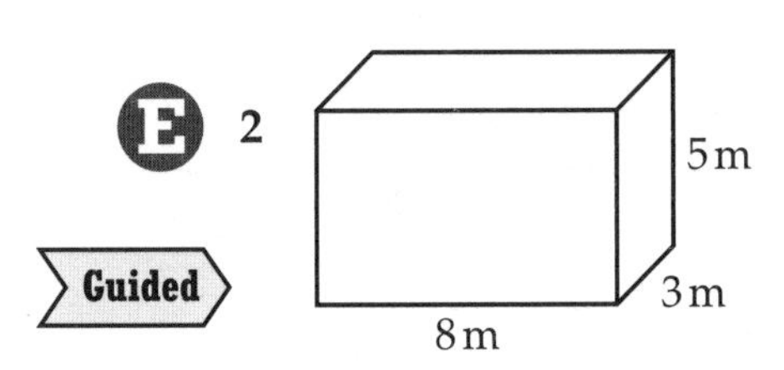

Here is a cuboid.
Work out the volume of the cuboid.

Volume of cuboid = length × width × height

= .... × .... × ....

= .... m^3

(2 marks)

 **3**

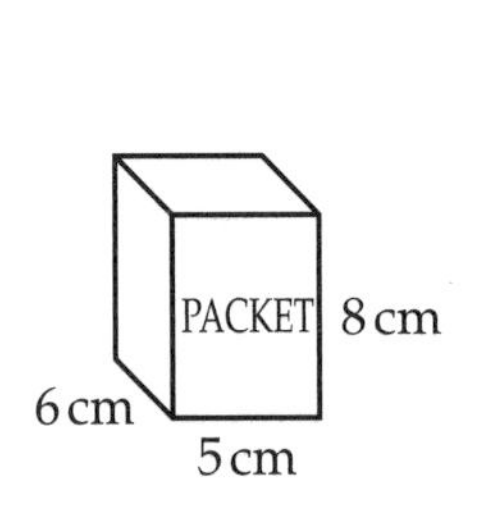

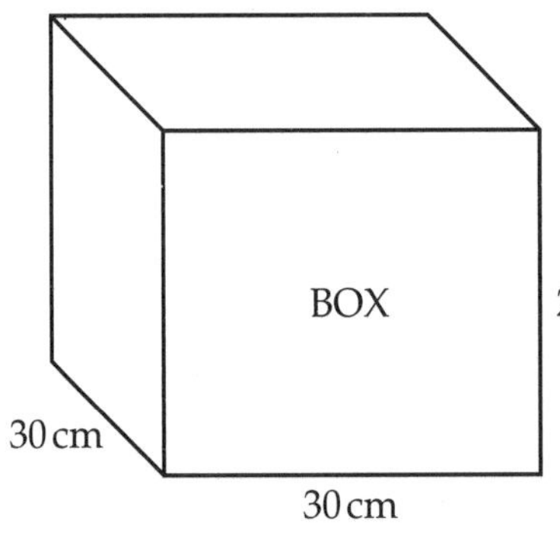

Diagram **NOT** accurately drawn

A packet measures 6 cm by 5 cm by 8 cm.
A box measures 30 cm by 30 cm by 24 cm.
The box is to be completely filled with packets.
Work out the number of packets which can completely fill the box.

First work out the volume of the box and the volume of the packet.

.... packets **(3 marks)**

Prisms

D **1** Work out the total surface area of this cuboid.

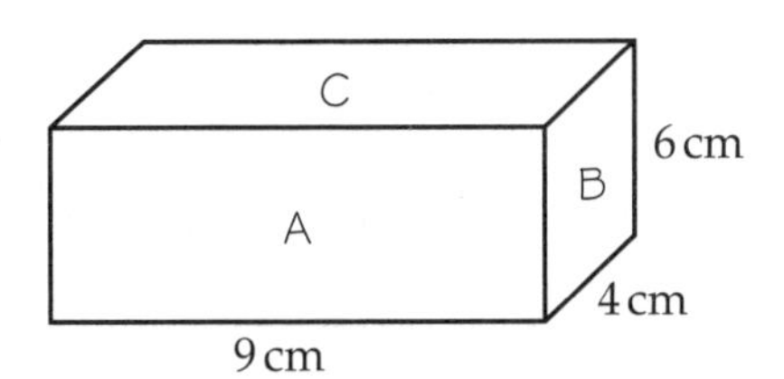

Guided

Total surface area = 2 × area of face A + 2 × area of face B + 2 × area of face C

= 2 ×......... × .... + 2 × .... × .... + 2 × .... × ....

= .... + ... + ...

= .... cm^2 **(2 marks)**

D **2** Work out the total surface area of this triangular prism.

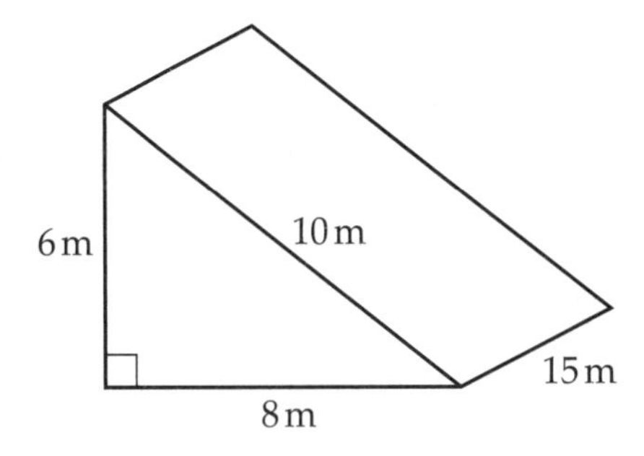

.... m^2 **(3 marks)**

D **3** Work out the total surface area of this L-shaped prism.

Check that you have included **all** the faces.

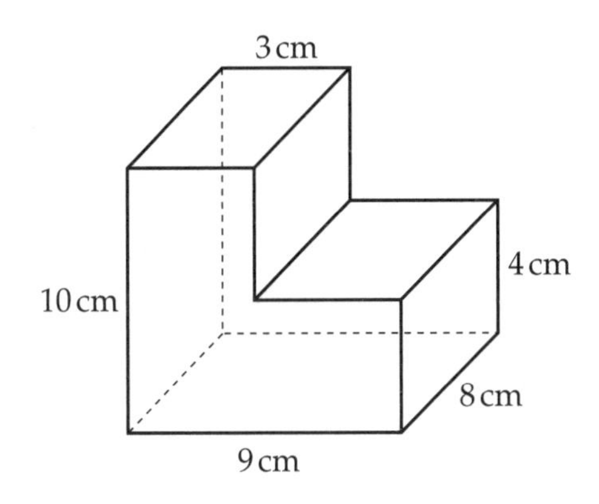

.... cm^2 **(4 marks)**

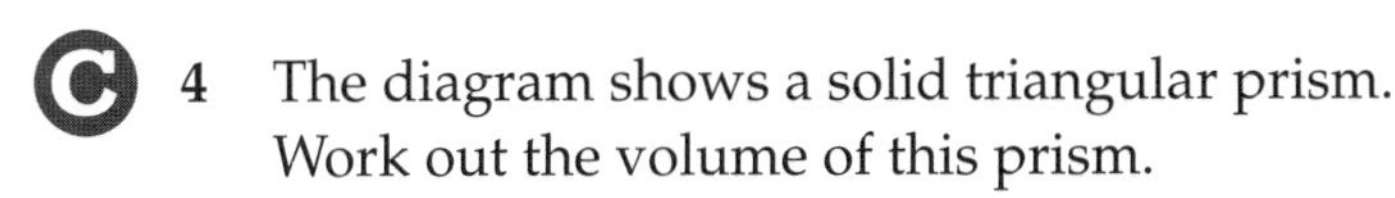

C **4** The diagram shows a solid triangular prism.
Work out the volume of this prism.

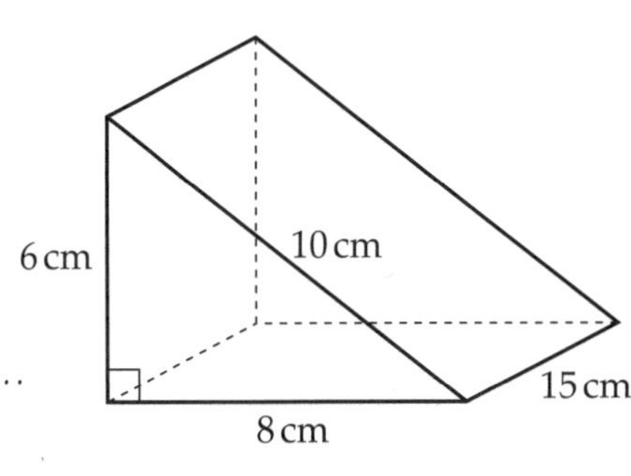

Guided

Volume of prism = area of cross-section × length

= area of triangle × length

= ($\frac{1}{2}$ × .... × ....) × ....

= .... × ....

= .... cm^3 **(3 marks)**

C **5** The area of the cross-section of this prism is 16 cm^2.
The length of the prism is 20 cm.
Work out the volume of the prism.

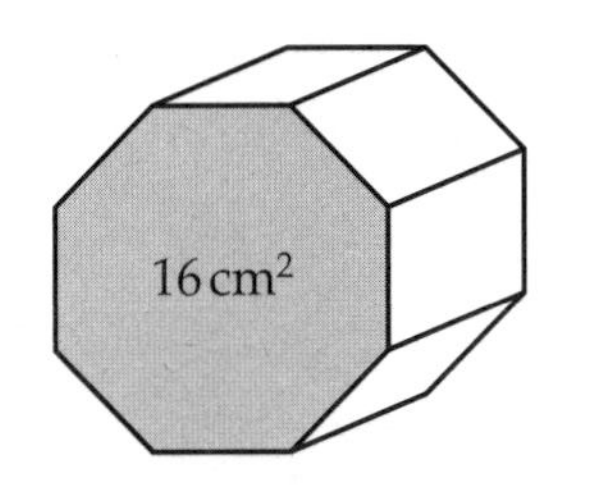

.... cm^3 **(2 marks)**

Cylinders

C **1** The diagram shows a cylinder of height 12 cm and radius 7 cm.
Work out the volume of the cylinder.
Give your answer correct to 3 significant figures.

7 cm
12 cm

Guided

Volume of cylinder = area of base × height
= area of circle × height
= $\pi r^2 h$
= π × ....^{2} × ....
= ....
= .... cm^3 to 3 s.f.

Write down at least six figures from your calculator.

(2 marks)

C **2** The diagram shows a cylinder with a height of 18 cm and a diameter of 12 cm.
Work out the volume of the cylinder.
Give your answer correct to 3 significant figures.

First work out the radius of the cylinder.

12 cm
18 cm

.... cm^3 **(2 marks)**

C **3** The diagram shows a solid cylinder.
The cylinder has a height of 20 cm and a radius of 14 cm.
Work out the **total** surface area of the cylinder.
Give your answer correct to 3 significant figures.

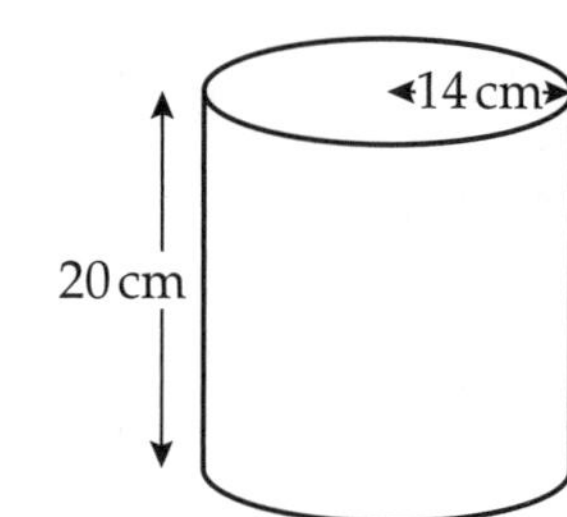

Guided

Total surface area = 2 × area of circle + curved surface area
= $2 \times \pi r^2 + 2\pi rh$
= 2 × π × ....^{2} + 2 × π × .... × ....
= .... + ...
= ....
= .... cm^2 to 3 s.f.

(3 marks)

C **4** A large water tank is in the shape of a cylinder.
The water tank has a height of 3 m and a radius of 2.5 m.
The water tank does not have a lid.
Jack is going to paint the outside of the water tank.
One tin of paint will cover 24 m^2.
How many tins of paint does Jack need to buy?

First work out the area of the curved surface and one circle.

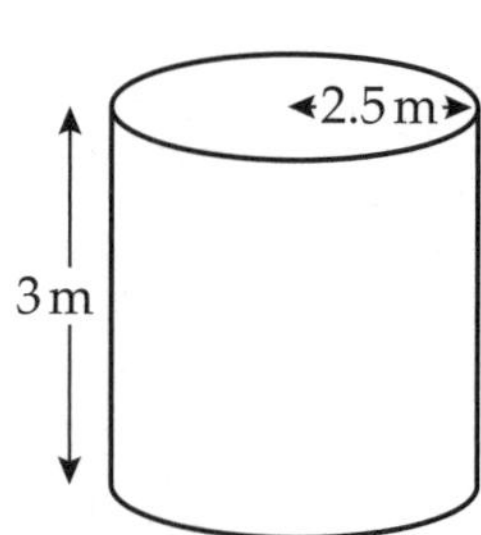

.... **(5 marks)**

Units of area and volume

C **1** Change 7.5 m^2 into cm^2.

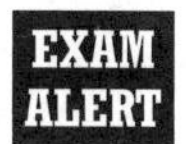

Guided

1 m = ... cm

1 m^2 = ...^{2} cm^2

= .... cm^2

7.5 m^2 = 7.5 × .... = ... cm^2 **(2 marks)**

C **2** Change 7 cm^3 into mm^3.

Exam questions similar to this have proved especially tricky – be prepared! ResultsPlus

Guided

EXAM ALERT

1 cm = 10 mm

1 cm^3 = ...^{3} mm^3

= .... mm^3

7 cm^3 = 7 × .... = ... mm^3 **(2 marks)**

C **3** Change 4 200 000 cm^3 into m^3.

You are changing a smaller unit to a larger unit so divide by a power of 10.

.... m^3 **(2 marks)**

C **4** Change 7850 mm^2 into cm^2.

.... cm^2 **(2 marks)**

C **5** Change 0.7 km^2 into m^2.

.... m^2 **(2 marks)**

C **6** The diagram shows a tank in the shape of a cuboid.
Work out how many litres of water the tank can hold.
1 litre = 1000 cm^3

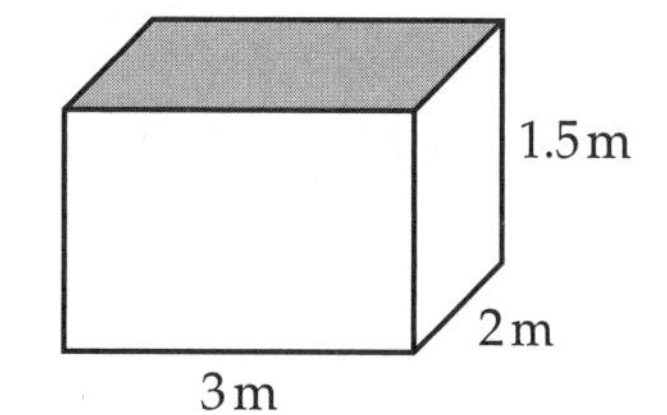

Diagram **NOT** accurately drawn

First change all the dimensions of the tank to centimetres.

.... litres **(4 marks)**

Constructions 1

1 Use ruler and compasses to **construct** the perpendicular to the line segment AB that passes through the point P.
You must show all construction lines.

For all constructions do **not** rub out the arcs you make when using your compasses.

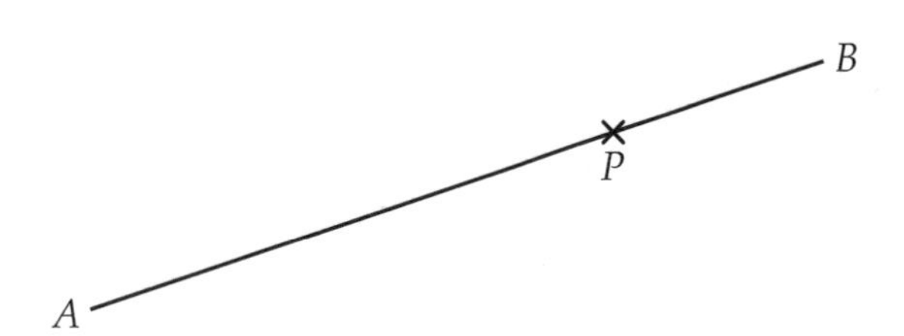

(2 marks)

2 Use ruler and compasses to construct the perpendicular bisector of the line AB.
You must show all your construction lines.

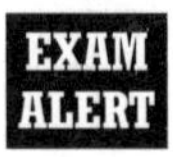

Exam questions similar to this have proved especially tricky – be prepared! ResultsPlus

A ——— B

(2 marks)

3 Use ruler and compass to construct the perpendicular from P to the line segment AB.
You must show **all** construction lines.

×P

A ——— B

(2 marks)

Constructions 2

D **1**

A

8 cm 6 cm

B 10 cm C

Diagram **NOT** accurately drawn

Use ruler and compasses to construct an accurate drawing of triangle ABC.
The line BC has been drawn for you.
You must show all your construction lines.

For all constructions do **not** rub out the arcs you make when using your compasses.

(2 marks)

C **2** Use ruler and compasses only to construct an angle of 45°.
You must show all your construction lines.

(4 marks)

Loci

C 1 ABC is a triangle.
Shade the region inside the triangle which is **both**
less than 5 cm from the point B
and
closer to the line AB than the line AC.

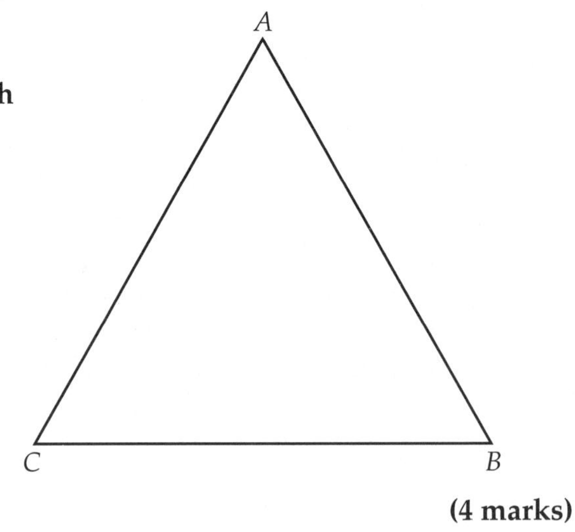

The question does **not** use the word construct so you can bisect angle *CAB* by using a protractor to measure the angle.

(4 marks)

C 2 The diagram shows two points, P and Q.

P ×   × Q

On the diagram shade the region that contains all the points that satisfy **both** the following:
the distance from P is less than 4 cm
the distance from P is greater than the distance from Q.

(4 marks)

C 3 $ABCD$ is a rectangle.
Shade the set of points inside the rectangle which are **both**
more than 2 cm from the line AB
and
more than 3.5 cm from the point C.

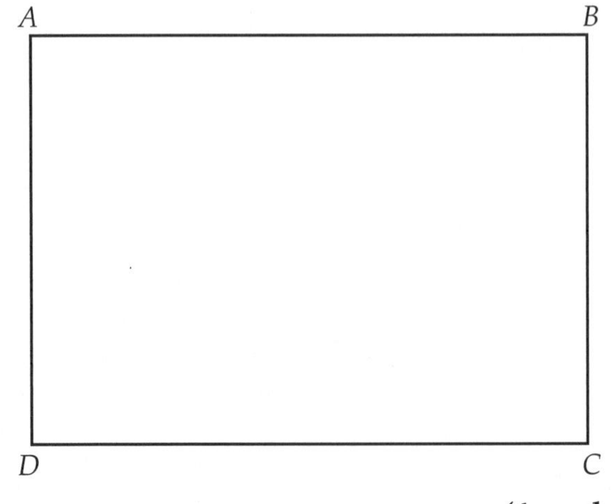

(4 marks)

Translations

C **1** (a) On the grid, translate triangle **A** by $\begin{pmatrix}5\\1\end{pmatrix}$.
Label the new triangle **B**.

Guided

$\begin{pmatrix}5\\1\end{pmatrix}$ means 5 units to the right and 1 unit up.

(2 marks)

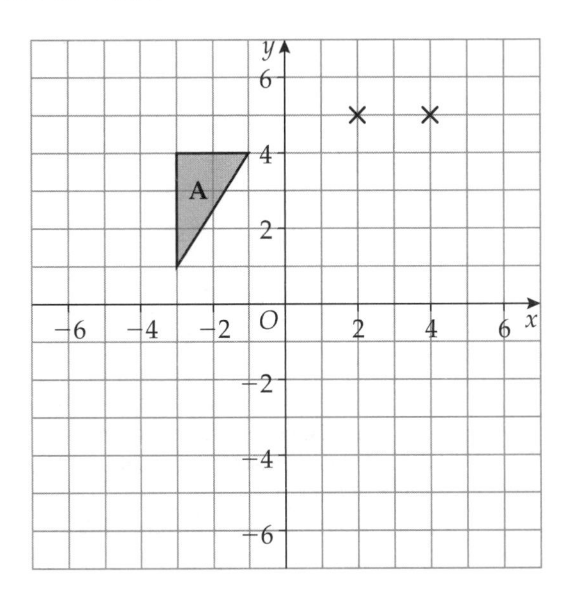

(b) On the grid, translate triangle **A** by $\begin{pmatrix}4\\-3\end{pmatrix}$.
Label the new triangle **C**.

$\begin{pmatrix}4\\-3\end{pmatrix}$ means 4 units to the right and 3 units down.

(2 marks)

C **2** On the grid, translate shape **D** by $\begin{pmatrix}-5\\-3\end{pmatrix}$.

(2 marks)

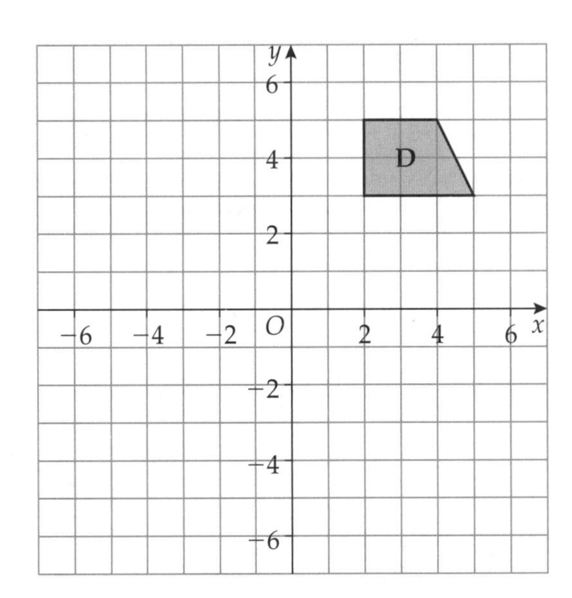

C **3** (a) Describe fully the single transformation that will map triangle **A** onto triangle **B**.

EXAM ALERT

....

....

(2 marks)

Exam questions similar to this have proved especially tricky – be prepared! ResultsPlus

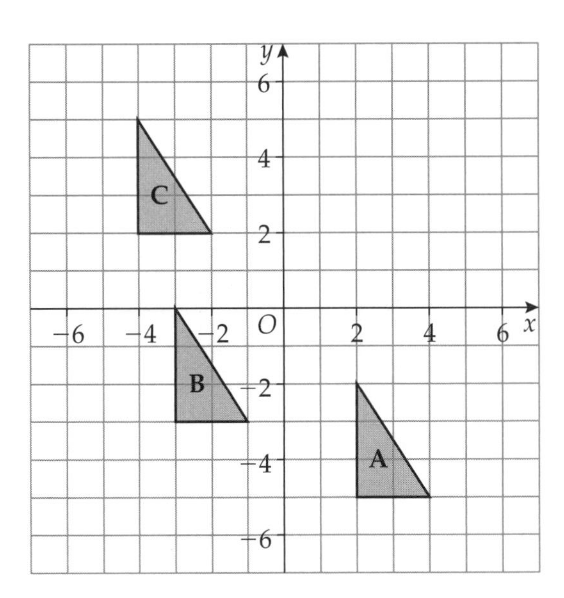

(b) Describe fully the single transformation that will map triangle **C** onto triangle **A**.

....

....

(2 marks)

Reflections

F **1** Reflect the shaded shape in the mirror line.

(a)

mirror line

Make sure that the mirror line is a line of symmetry for your completed answer.

(1 mark)

(b)

mirror line

The reflected shape should be the same distance from the mirror line as the original shape.

(2 marks)

C **2** (a) On the grid, reflect triangle **A** in the line $x = 3$
Label the new triangle **B**.

First draw the line $x = 3$ on the graph.

(2 marks)

(b) On the grid, reflect triangle **A** in the line $y = -x$
Label the new triangle **C**.

Turn your diagram 45° clockwise so that the mirror line is vertical.

(2 marks)

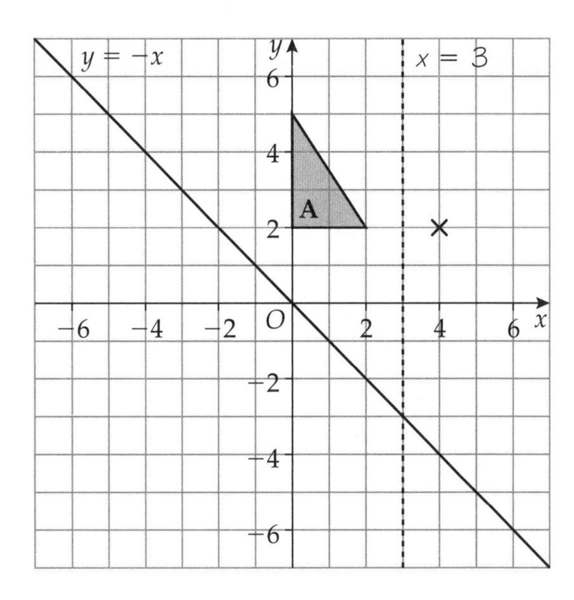

C **3** (a) Describe fully the single transformation that will map triangle **Q** onto triangle **T**.

....

....

(2 marks)

(b) Describe fully the single transformation that will map triangle **Q** onto triangle **P**.

....

....

(3 marks)

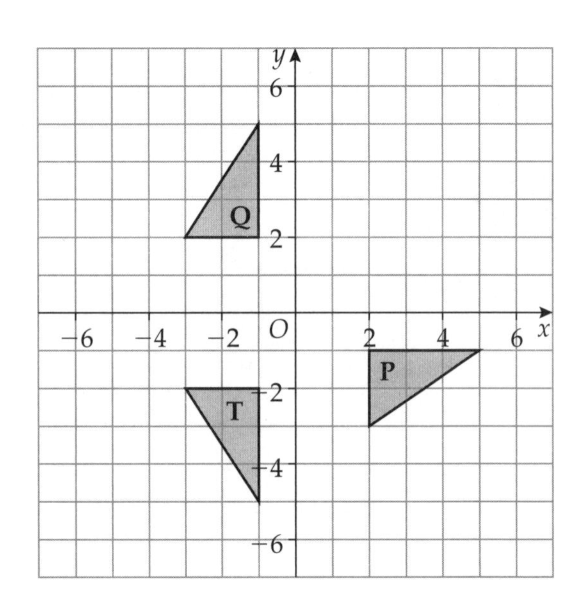

Rotations

C 1 On the grid, rotate triangle **A** 180° about O (0, 0).

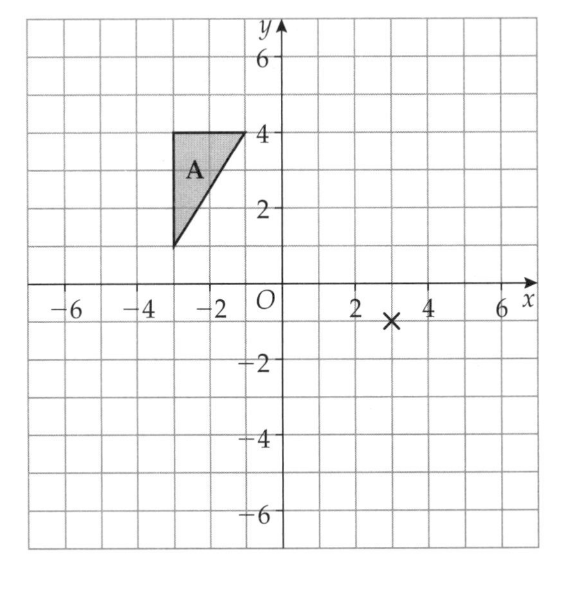

Guided

> Trace over triangle **A** using tracing paper. Put your pencil at the point (0, 0) and then rotate the tracing paper through 180°.

(2 marks)

C 2 On the grid, rotate shape **D** 90° anticlockwise about (1, 1). **(2 marks)**

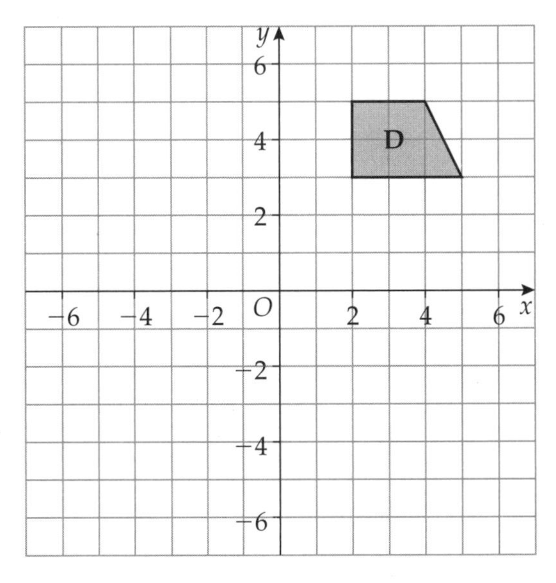

C 3 (a) Describe fully the single transformation that will map triangle **P** onto triangle **T**.

....

....

(3 marks)

(b) Describe fully the single transformation that will map triangle **P** onto triangle **Q**.

....

....

(3 marks)

Enlargements

D **1** On the grid, enlarge the shape with a scale factor of 3

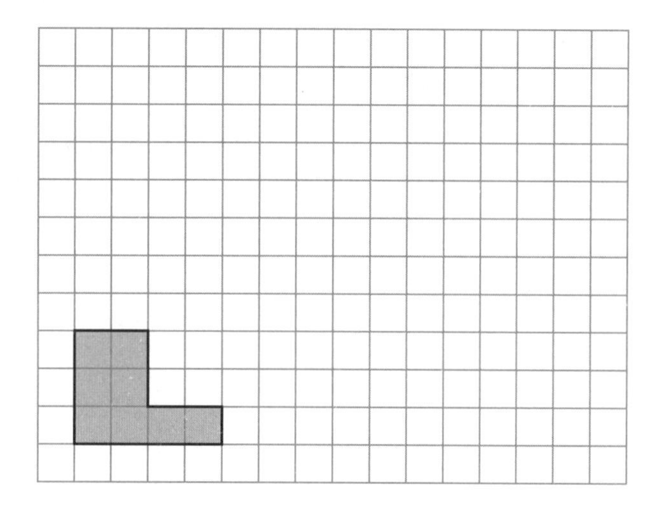

No centre of enlargement is given so the enlarged shape can be placed anywhere on the grid.

(2 marks)

D **2** On the grid, enlarge the shape with a scale factor of 2, centre (0, 0).

(3 marks)

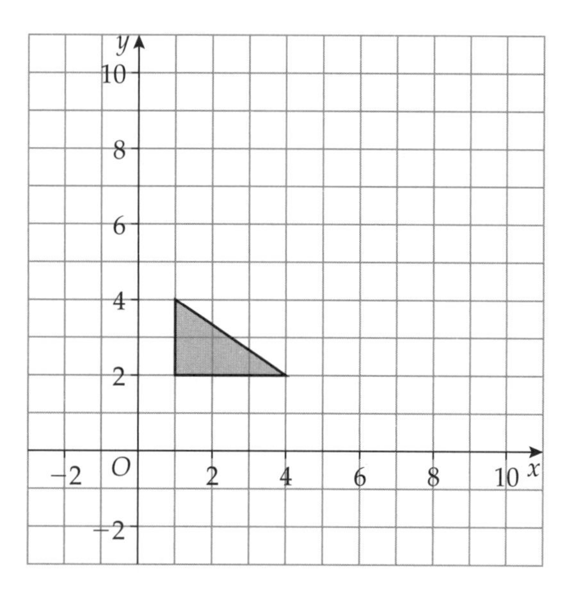

C **3** Describe fully the single transformation that maps triangle **A** onto triangle **B**.

....

....

(3 marks)

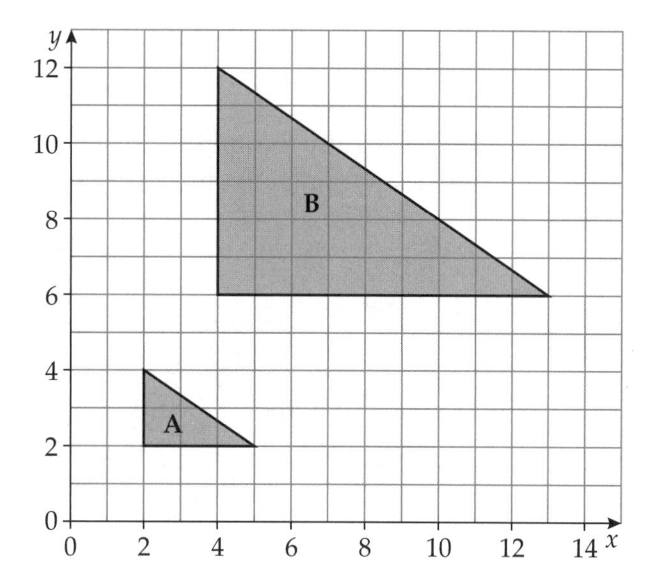

Combining transformations

C 1 (a) Reflect triangle **A** in the x-axis.
Label the new triangle **B**.

(2 marks)

(b) Reflect triangle **B** in the y-axis.
Label the new triangle **C**.

(2 marks)

(c) Describe fully the single transformation that will map triangle **A** onto triangle **C**.

....

....

(1 mark)

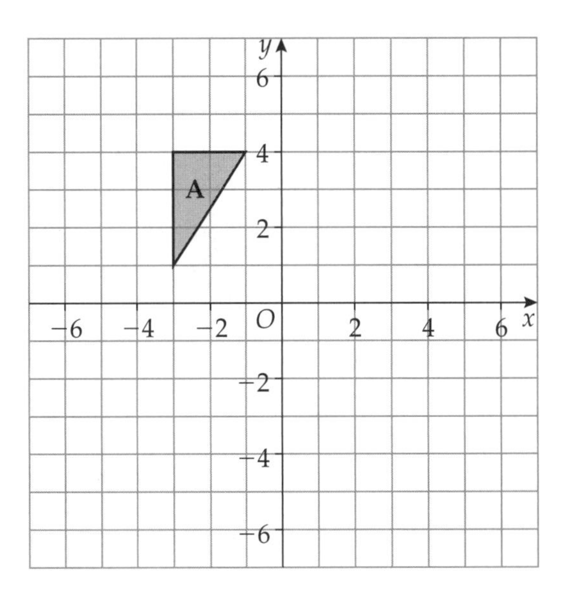

C 2 (a) Reflect shape **D** in the line $y = -3$
Label the new shape **E**.

(2 marks)

(b) Reflect shape **E** in the x-axis.
Label the new shape **F**.

(2 marks)

(c) Describe fully the single transformation that will map shape **D** onto shape **F**.

....

....

(1 mark)

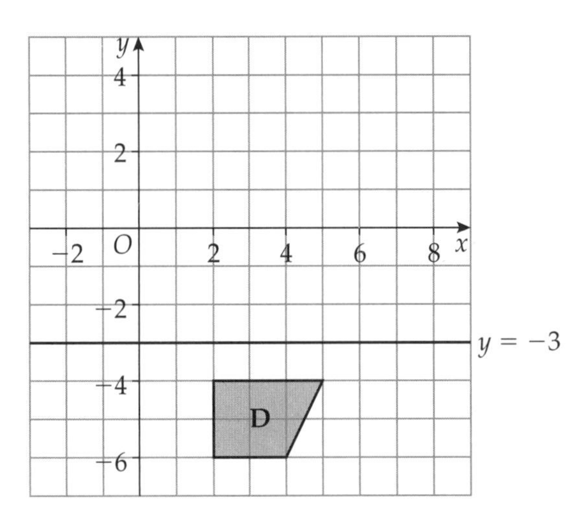

C 3 (a) Translate **P** by $\begin{pmatrix} -5 \\ -7 \end{pmatrix}$.
Label the new triangle **Q**.

(2 marks)

(b) Translate **Q** by $\begin{pmatrix} 1 \\ 4 \end{pmatrix}$.
Label the new triangle **R**.

(2 marks)

(c) Describe fully the single transformation that will map triangle **P** onto triangle **R**.

....

....

(1 mark)

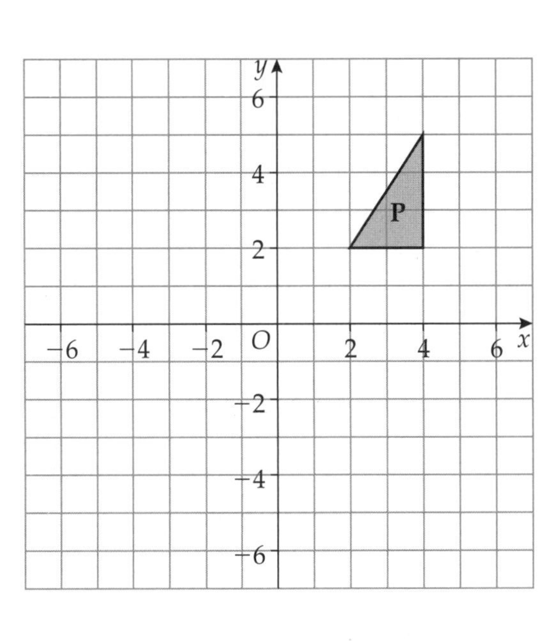

Similar shapes

1 Here are seven triangles on a square grid.

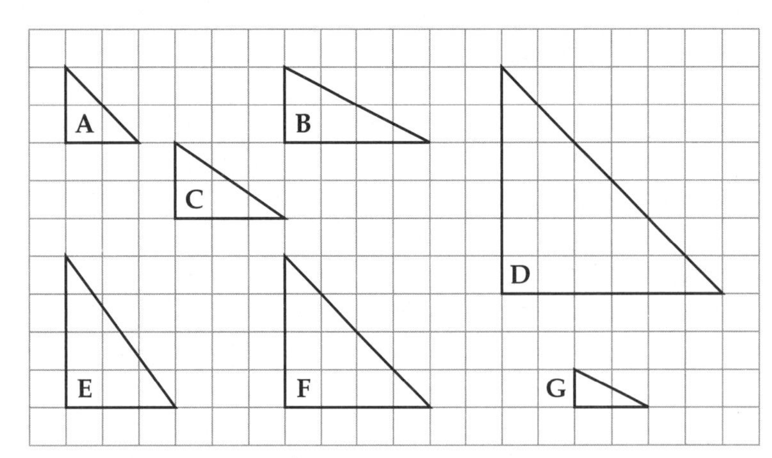

If two triangles are similar, then one is an enlargement of the other.

Two of the triangles are similar to triangle **A**.

(a) Write down the letters of these two triangles.

.... and .... **(2 marks)**

One of the triangles is similar to triangle **G**.

(b) Write down the letter of this triangle.

.... **(1 mark)**

2 Here are two similar shapes.

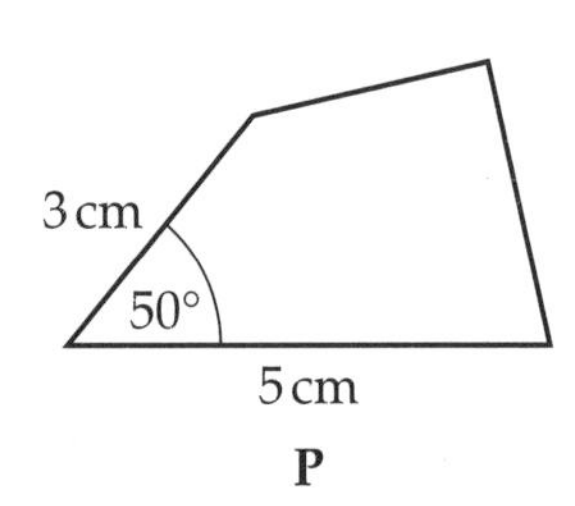

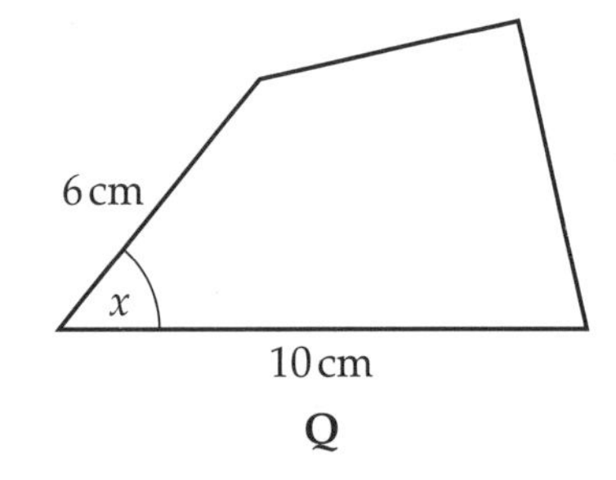

Diagram **NOT** accurately drawn

Shape **P** is similar to shape **Q**.
Write down the size of the angle marked x.

.....................° **(1 mark)**

3 Rectangle **P** has a length of 4 cm.
Rectangle **P** has a width of 2 cm.

Diagram **NOT** accurately drawn

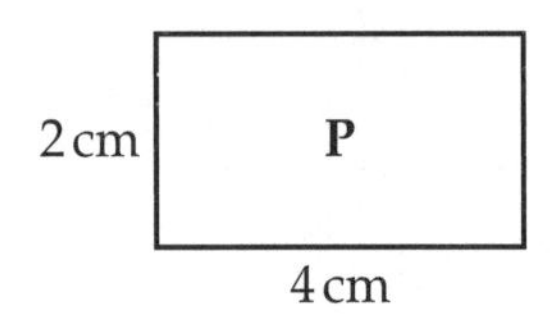

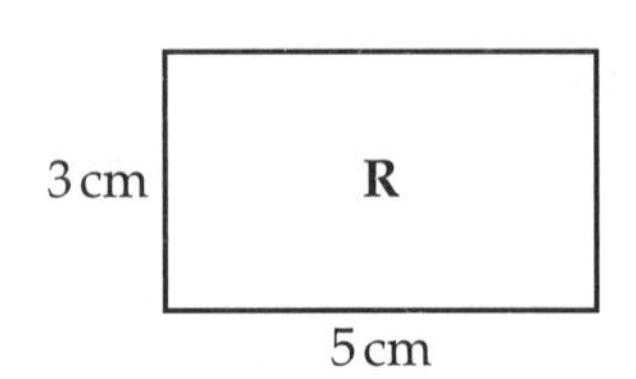

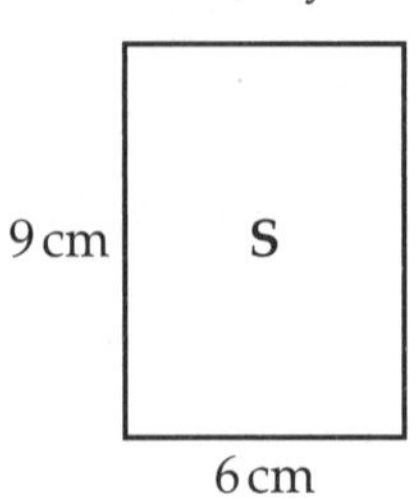

Which of the above shapes is similar to rectangle **P**?

.... **(1 mark)**

Pythagoras' theorem

C **1** Work out the length of PQ.
Give your answer correct to 3 significant figures.

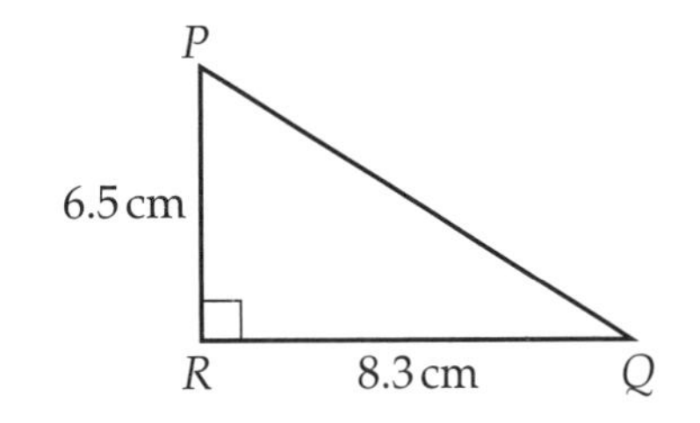

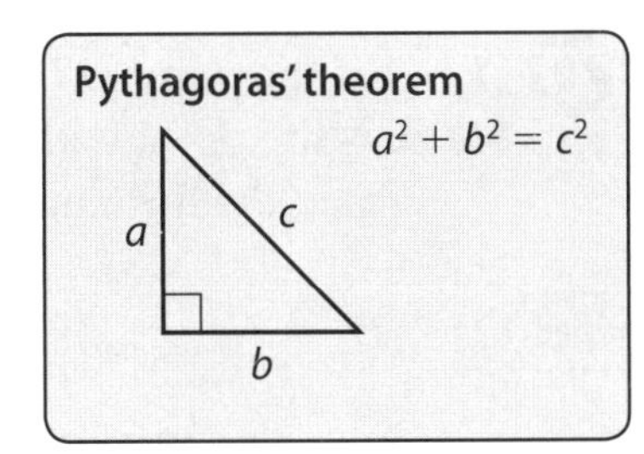

Guided

$PQ^2 = 6.5^2 + \dots\dots\dots\dots$

$PQ^2 = \dots\dots\dots\dots$

$PQ = \sqrt{\dots\dots\dots\dots}$

$PQ = \dots\dots\dots\dots\dots\dots\dots$

$PQ = \dots\dots\dots\dots$ cm to 3 s.f.

(3 marks)

C **2** Work out the length of AB.
Give your answer correct to 3 significant figures.

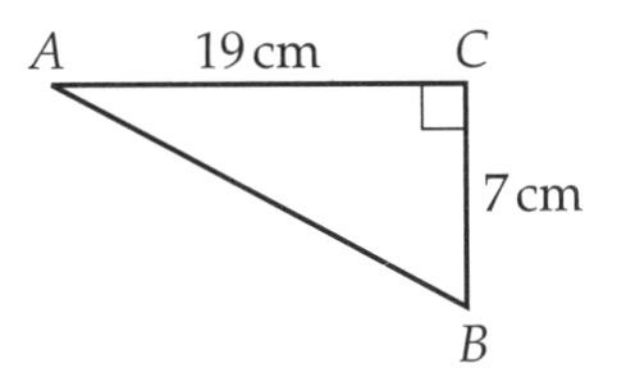

.... cm **(3 marks)**

C **3** Work out the length of DE.

$DE^2 + \dots\dots\dots\dots = 26^2$

$DE^2 = 26^2 - \dots\dots\dots\dots$

$DE^2 = \dots\dots\dots\dots$

$DE = \sqrt{\dots\dots\dots\dots}$

$DE = \dots\dots\dots\dots$ m

The hypotenuse is 26 m so you subtract rather than add because you are finding one of the shorter sides.

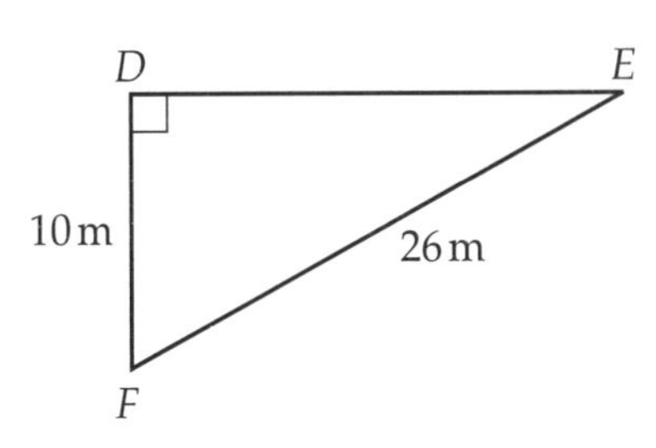

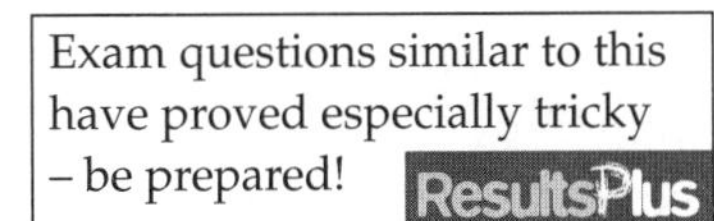

(3 marks)

C **4** Work out the length of LM.
Give your answer correct to 3 significant figures.

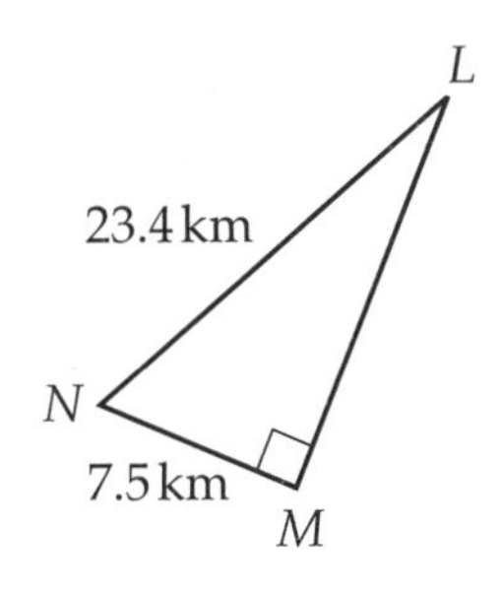

.... km **(3 marks)**

Line segments

D **1** A is the point with coordinates (1, 2).
B is the point with coordinates (13, 7).
Calculate the length of AB.

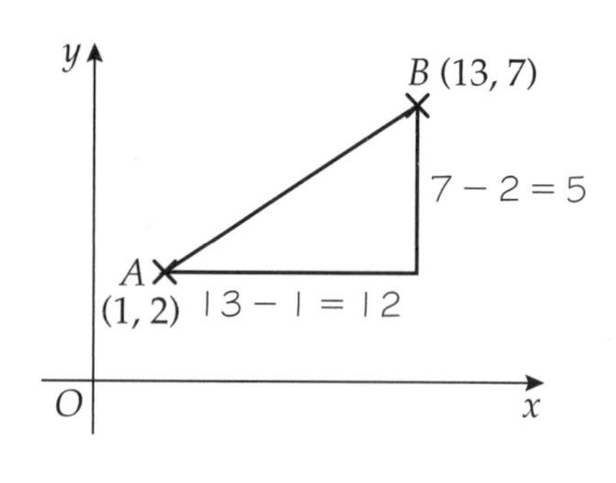

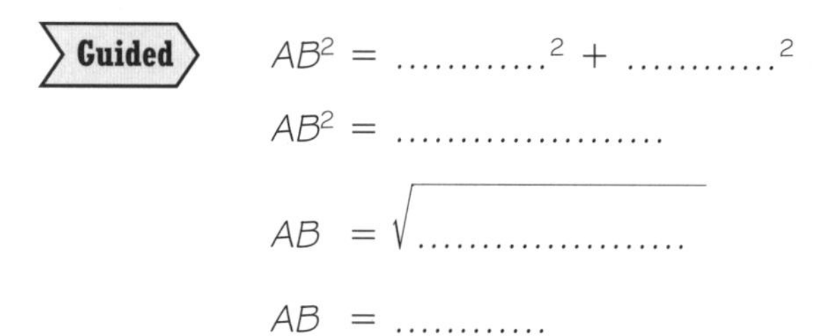

AB^2 = ...^{2} + ...^{2}

AB^2 = ...

AB = $\sqrt{......................}$

AB = ...

(3 marks)

D **2** C is the point with coordinates (5, 9).
D is the point with coordinates (2, 4).
Calculate the length of CD.

Sketch a diagram like that in question 1.

CD = ... **(3 marks)**

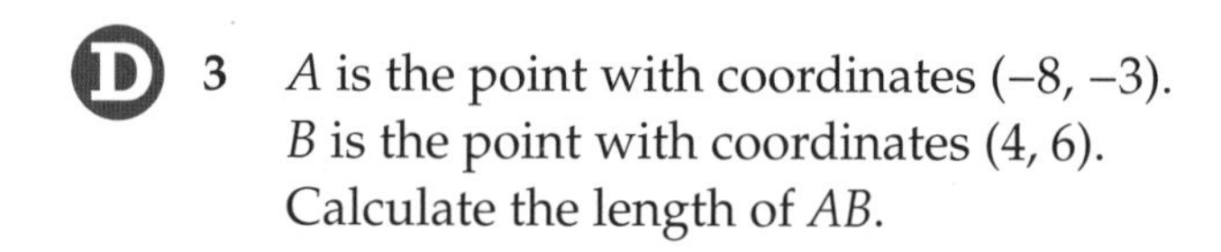

D **3** A is the point with coordinates (−8, −3).
B is the point with coordinates (4, 6).
Calculate the length of AB.

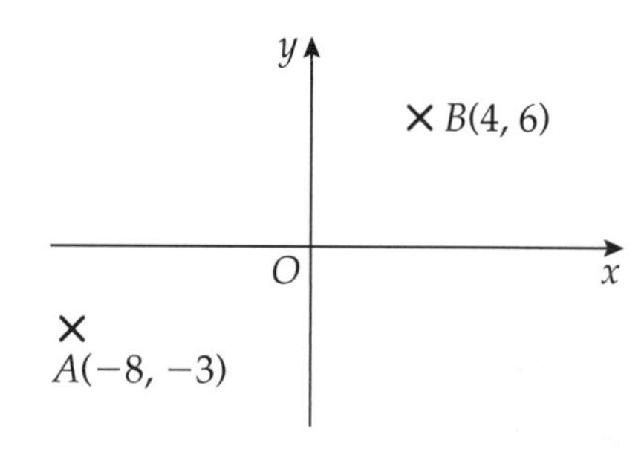

AB = ... **(3 marks)**

C **4** L is the point with coordinates (−9, 10).
M is the point with coordinates (1, −14).
Calculate the length of LM.

LM = ... **(3 marks)**

Problem-solving practice

E **1** On the grid, show how this shape will tessellate.
You should draw at least six shapes.

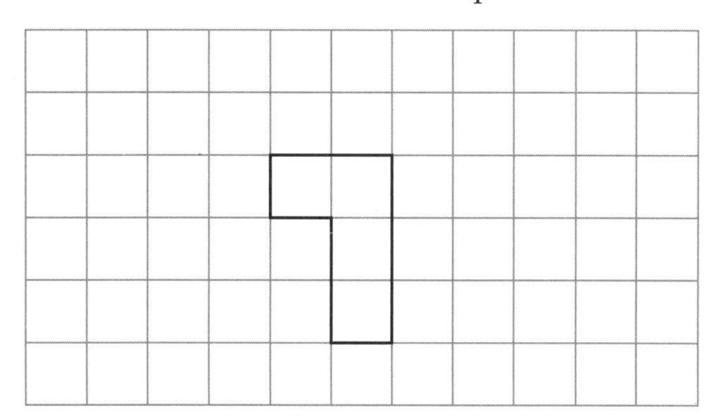

Make sure that each shape is exactly the same size. There should be no gaps or overlaps.

(2 marks)

D **2** The diagram shows a prism.
Work out the volume of the prism.

Split the prism into two cuboids.
Remember to include units with your answer.

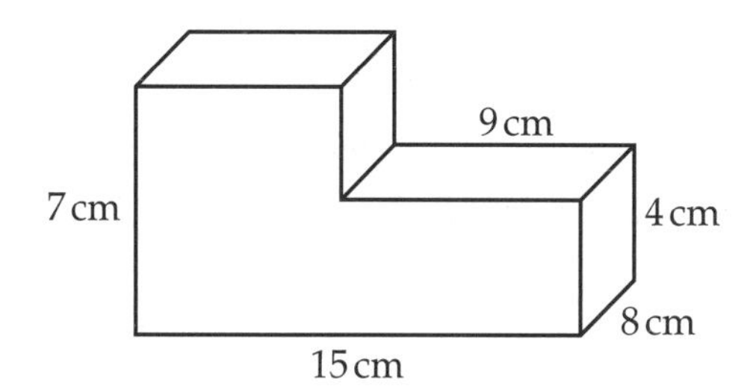

.... **(4 marks)**

C **3** ABC is parallel to $DEFG$.
$BF = EF$
Angle $CBF = 42°$

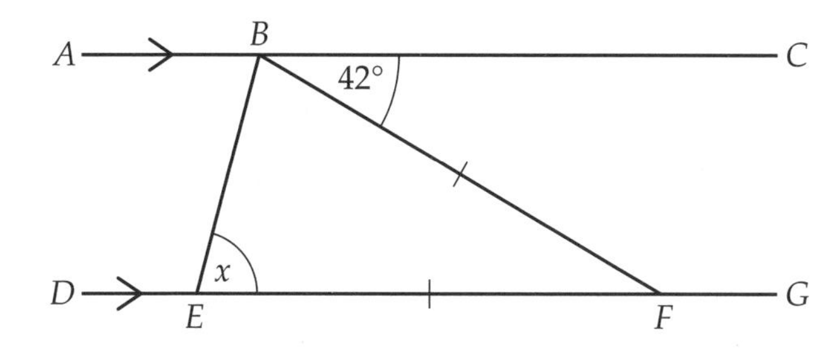

Diagram **NOT** accurately drawn

Work out the value of x.
You must give a reason for each stage in your working.

Write down each angle fact that you use.

Guided

Angle BFE = ...° .... are equal.

x = (180° − ....) ÷ 2 Angles in a triangle ....

= .... ÷ 2 Base angles of an ....

= ....°

(4 marks)

Problem-solving practice

4 Nathalie tries to make a tessellation using just regular octagons.
Explain why she cannot tessellate with just octagons.

Guided

Exterior angle of a regular octagon = 360 ÷ ....

= ....°

Interior angle of a regular octagon = 180 − ....

= ....°

Number of octagons around a point = 360 ÷ ....

= ....

Octagons will not tessellate because ....

....

(3 marks)

*5 Jim has a fishpond in his garden.
The fishpond is in the shape of a circle.
He wants to put fencing around the edge of his fishpond.
The fishpond has a diameter of 3.6 metres.
Fencing costs £5.69 per metre.
Jim can only buy whole metres of fencing.
Work out the total cost of the fencing that Jim needs to buy.

The question has a * next to it, so you must show all your working and write your answer clearly with the correct units in a sentence.

Work out the circumference of the pond. Decide how many whole metres of fencing Jim will need and work out the total cost.

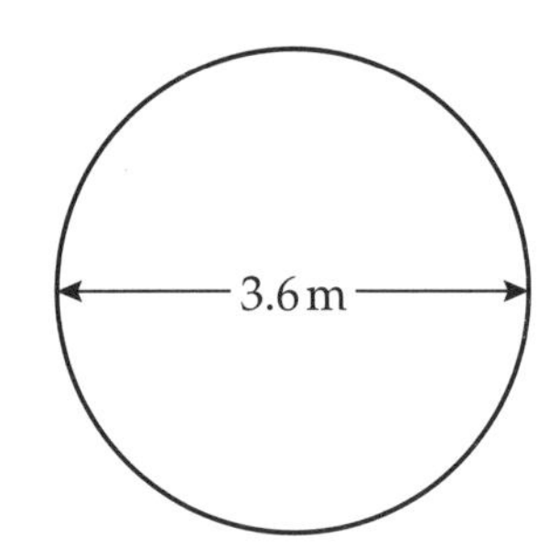

(4 marks)

6 Describe fully the single transformation that will map triangle **A** onto triangle **B**.

....

....

....

There are 3 marks for the question so you should give 3 pieces of information.

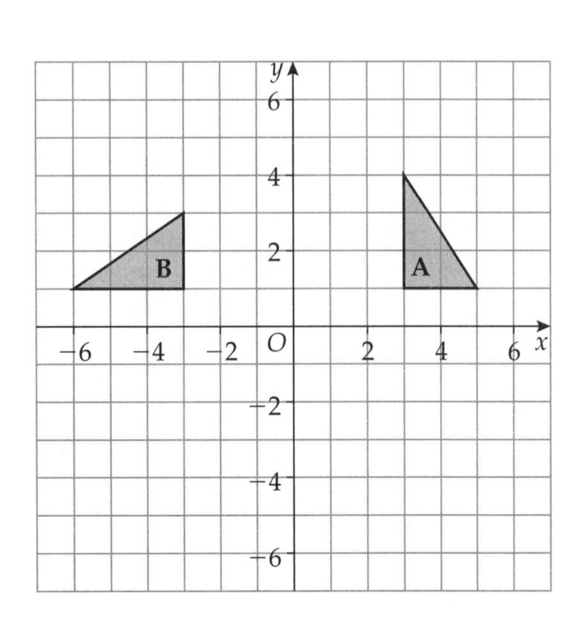

(3 marks)

Collecting data

D **1** Mary wants to find out people's favourite pet.
Design a data collection sheet she could use to find this out.

Guided

EXAM ALERT

Pet		

Exam questions similar to this have proved especially tricky – be prepared! ResultsPlus

(2 marks)

D **2** Sam wants to collect information about the different ways people travel to work.
Design a suitable data collection table that Sam could use to collect this information.

(3 marks)

D **3** Katie wants to find out how often her friends go to the cinema.
She uses this question on a questionnaire.

How many times do you go to the cinema?		
☐ 1–2	☐ 2–5	☐ 6+

Make sure that you find different things that are wrong – look at both the question and the response boxes.

Write down **two** things wrong with this question.

1 ....

2 ....

(2 marks)

C **4** Anthony wants to design a question for a questionnaire to find out how often people go on holiday.
Design a suitable question he could use.

Remember to include a time frame and some response boxes with your question.

(2 marks)

C **5** Lola wants to design a question for a questionnaire to find out how far people travel to work.
Design a suitable question she could use.

(2 marks)

Two-way tables

D 1 80 British students each visited one foreign country last week.
The two-way table shows some information about these students.

	France	Germany	Spain	Total
Female			13	38
Male	21			
Total		25	22	80

Work out the missing values in this row first.
Start with the total column.

Complete the two-way table.

Guided 'Total' column: 80 − 38 = 42

(3 marks)

D 2 The two-way table shows some information about the lunch arrangements of 60 students.

	School lunch	Packed lunch	Other	Total
Female	16			39
Male	9	8		
Total			21	60

Complete the two-way table. **(3 marks)**

C 3 40 students were asked if they liked coffee.
25 of the students were girls.
13 boys liked coffee.
14 girls did **not** like coffee.
Use this information to complete the two-way table.

Transfer the information from the question to the two-way table first. Then complete the table.

	Boys	Girls	Total
Liked coffee	....		
Did not like coffee		....	
Total		....	40

(3 marks)

C 4 75 adults were asked which type of music they preferred.
47 of the adults were male.
2 of the females and 8 of the males preferred country.
32 of the adults preferred jazz.
30 of the males preferred jazz.
Draw up a two-way table to show this information.

(4 marks)

Pictograms

G **1** The pictogram shows the numbers of hours of sunshine on Monday, Tuesday and Wednesday one week.

Monday	⦶ ⦶ ⦶
Tuesday	⦶ ⦶ ⦶ ⦶ ⦶ ⦶
Wednesday	⦶ ⦶ ⦶ ◖
Thursday	
Friday	

Key: ⦶ represents 2 hours

(a) Write down the number of hours of sunshine

(i) on Tuesday (ii) on Wednesday.

Guided

Number of full circles on Tuesday = ...

Number of hours of sunshine on Tuesday = ... × 2

= ....

.... hours **(2 marks)**

On Thursday there were 6 hours of sunshine.

(b) Show this on the pictogram.

Draw in the circles in the space next to Thursday.

Guided

Number of circles = 6 ÷ 2 = ... **(1 mark)**

On Friday there were 9 hours of sunshine.

(c) Show this on the pictogram.

(1 mark)

2 Barry buys DVDs. The pictogram shows information about the number of comedy DVDs and the number of action DVDs he buys.

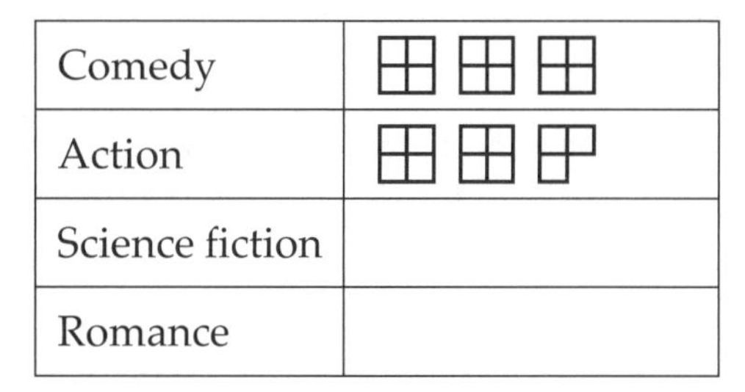

Comedy	⊞ ⊞ ⊞
Action	⊞ ⊞ (3 small squares)
Science fiction	
Romance	

Key: ⊞ represents 8 DVDs

(a) Write down the number of comedy DVDs he buys.

.... **(1 mark)**

(b) Write down the number of action DVDs he buys.

Each small square represents 8 ÷ 4 = 2

.... **(1 mark)**

Barry buys 32 science fiction DVDs.

(c) Show this information on the pictogram.

(1 mark)

Barry buys 10 romance DVDs.

(d) Show this information on the pictogram.

(1 mark)

Bar charts

G **1** The bar chart shows information about the way the students in class 7X travelled to school.

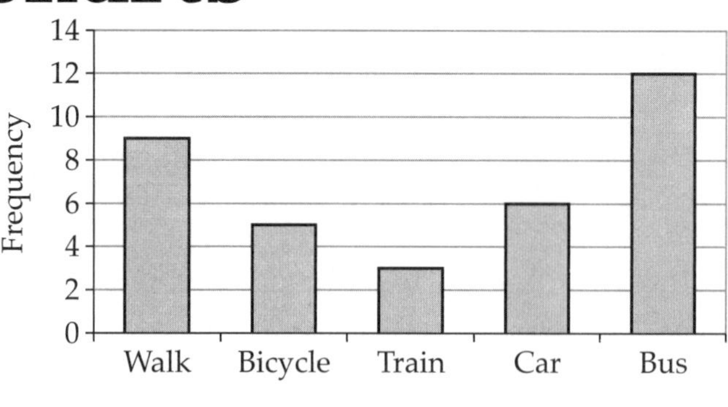

(a) How many students travelled by train?

.... **(1 mark)**

(b) Which was the most popular way of travelling to school?

.... **(1 mark)**

(c) How many students are there in class 7X?

Find the height of each bar then add up all the heights.

.... **(2 marks)**

G **2** The bar chart shows information about the amounts of time Alex and Mark spent playing computer games on each of four days last week.

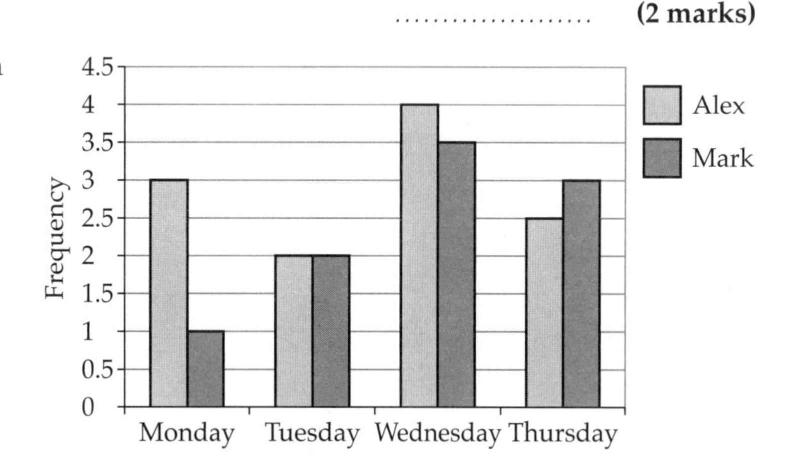

(a) Write down the number of hours that Alex played computer games on Thursday.

.... **(1 mark)**

(b) On which day did Alex and Mark play computer games for the same amount of time?

.... **(1 mark)**

F **3** Sally asked 24 of her friends to name their favourite colour.

~~blue~~	~~blue~~	red	yellow	purple	red	yellow	purple	blue	blue	red	blue
yellow	purple	blue	yellow	blue	purple	yellow	red	red	blue	purple	red

(a) Use Sally's results to complete the tally chart.

Guided

	Tally	Frequency
Blue	||	
Red		
Yellow		
Purple		

(2 marks)

(b) Draw a chart or diagram to show this information.

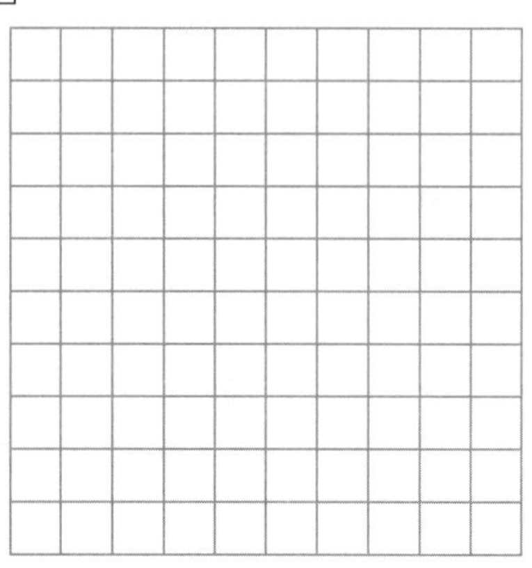

(3 marks)

Frequency polygons

1 Emily carried out a survey of 70 students.
She asked them how many DVDs they each have.
This table gives information about the numbers of DVDs these students have.

Midpoint	2	7	12	....	....
Number of DVDs	0–4	5–9	10–14	15–19	20–24
Frequency	6	19	15	23	7

On the grid, draw a frequency polygon to show this information.

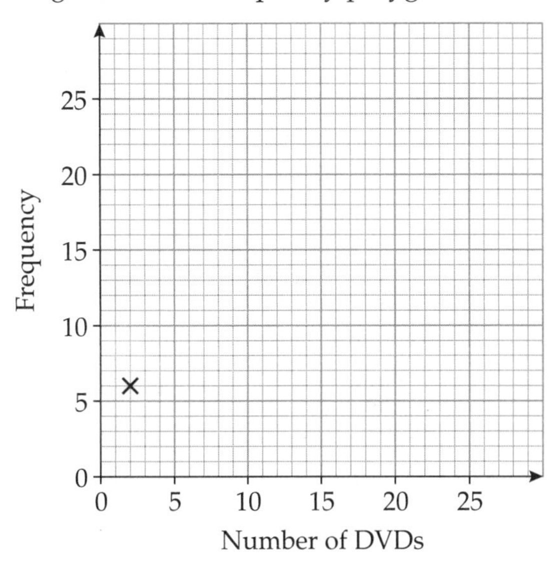

Plot the points at the midpoint of each class interval. Plot the second point at (7, 19).

Join up the plotted points with straight lines.

(2 marks)

2 The table gives some information about the ages, in years, of 50 people.

Age (years)	Frequency
0–9	4
10–19	8
20–29	14
30–39	12
40–49	9
50–59	3

On the grid, draw a frequency polygon to show this information.

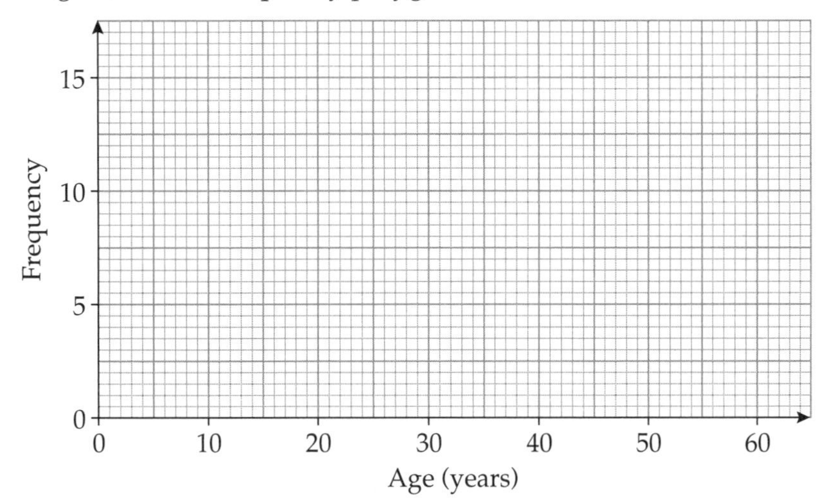

(2 marks)

Pie charts

E 1 Malik asked 60 of his friends which flavour of crisp they prefer.
The table shows his results.

Flavour of crisp	Frequency
Plain	14
Prawn cocktail	8
Cheese & Onion	18
Salt & Vinegar	20

Draw an accurate pie chart to show these results.

Guided

Angle for 1 packet of crisps = 360 ÷ 60

= ....°

Angle for Plain = 14 × .... = ...°

Angle for Prawn cocktail = ... × .... = ...°

Angle for Cheese & Onion = ... × .... = ...°

Angle for Salt & Vinegar = ... × .... = ...°

(3 marks)

Add up your four angles and check that the total is 360°. Now draw each angle carefully on the pie chart. Label the sectors Plain, Prawn cocktail, etc.

E 2 The pie chart gives information about the pets owned by students in Year 9.

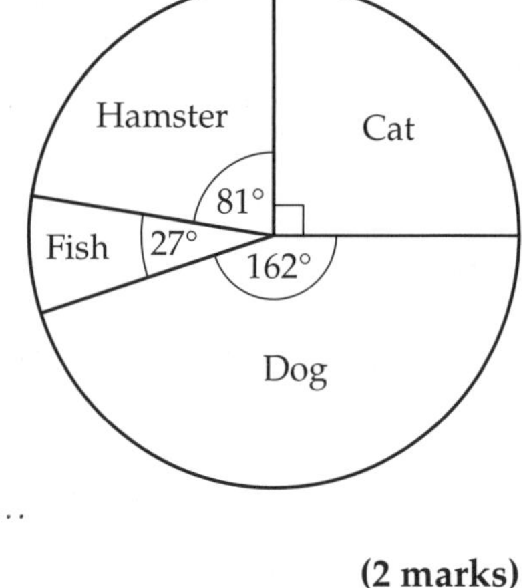

(a) Which pet was the most popular?

.... **(1 mark)**

10 students own a cat.

(b) How many students own a dog?

Guided

Number of degrees that represent 1 pet = 90 ÷ 10

= ....°

Number of students who own a dog = 162° ÷ ....

= ....

(2 marks)

(c) How many students own a hamster?

.... **(2 marks)**

D 3 The table gives information about the drinks sold in a café one day.

Drink	Frequency
Coffee	48
Tea	32
Orange juice	16
Hot chocolate	24

Draw an accurate pie chart to show these results.

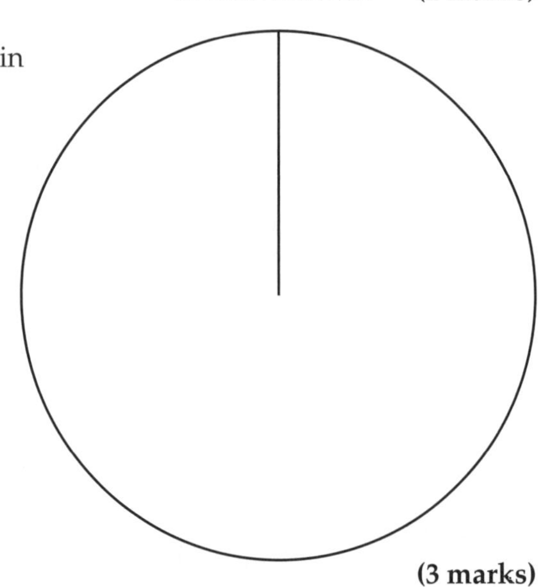

(3 marks)

Averages and range

F **1** Here are eight numbers.

1 1 2 3 4 5 5 5

(a) Write down the mode.

The mode is the value that appears most often.

Guided The mode is ....

(1 mark)

(b) Work out the median.

There are two middle numbers. Find the mean of these.

Guided Median = $\frac{\text{...........} + \text{...........}}{2}$

= ....

(1 mark)

(c) Work out the mean.

Guided The total of all values = 1 + 1 + 2 + 3 + 4 + 5 + 5 + 5

= ....

Mean = $\frac{\text{...........}}{8}$

= ....

(2 marks)

(d) Find the range.

Guided Range = largest value − smallest value

= .... − ....

= ....

(2 marks)

F **2** Here are the numbers of texts received by 10 adults in one day.

2 4 7 6 25 24 16 5 4 8

(a) Write down the mode.

....

(1 mark)

(b) Work out the median.

First write out all the numbers in order.

....

(2 marks)

(c) Work out the mean.

....

(2 marks)

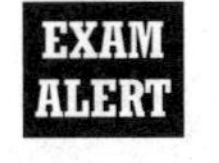

3 Lucas has three cards. Each card has a number on it.
The numbers are hidden.
The median of the three numbers is 6
The mean of the three numbers is 7
Work out which three numbers could be on the cards.

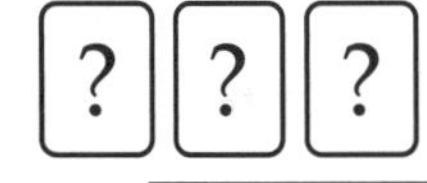

Exam questions similar to this have proved especially tricky – be prepared! ResultsPlus

The median of the three numbers is 6 so 6 must be the number on the .... card.

The total of the 3 numbers on the cards = 7 × 3 = 21

The numbers on the cards could be ...., 6, ...

(2 marks)

Stem and leaf diagrams

E 1 The stem and leaf diagram shows the numbers of minutes 19 students took to complete their maths homework.

0	7 8 8 9
1	2 3 5 6
2	4 4 5 5 8
3	0 0 0 5 7 8

Use the key to interpret the stem and leaf diagram.

Key: 3 | 0 stands for 30 minutes

(a) Work out the range.

EXAM ALERT

Range = highest value − lowest value

= 38 − ....

= ....

Exam questions similar to this have proved especially tricky – be prepared! ResultsPlus

(2 marks)

(b) Find the median.

The median will be the middle number in the stem and leaf diagram.

.... minutes **(1 mark)**

E 2 Jake counted the number of letters in each of 20 sentences in a newspaper. He showed his results in a stem and leaf diagram.

0	6 6 7 8
1	1 2 3 4 4 8 9
2	0 3 5 5 5 8
3	2 2 3

Key: 0 | 6 represents 6 letters

(a) Write down the mode.

.... **(1 mark)**

(b) Find the median.

.... **(1 mark)**

(c) Work out the range.

.... **(2 marks)**

D 3 Here are the ages of 20 people.

25 14 9 27 29 30 33 45 41 7 38 40 49 21 25 26 28 18 19 19

In the space below, draw an ordered stem and leaf diagram to show this information.

Guided

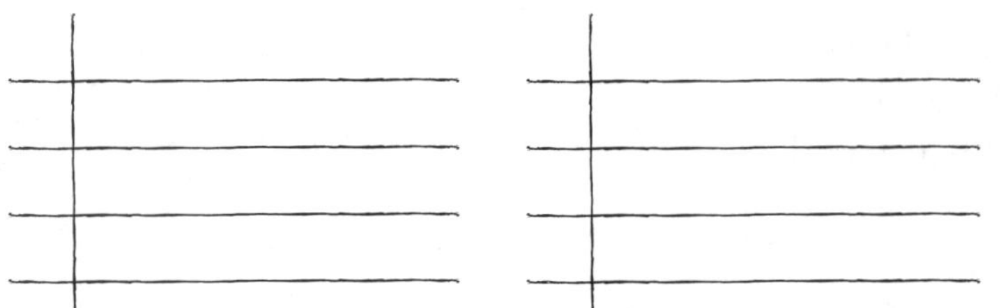

Start by drawing a suitable grid. Remember to include a key.

Key:

(3 marks)

Averages from tables 1

1 The table shows information about the numbers of goals scored by a football team in their last 20 matches.

Guided

Number of goals	Frequency	Number of goals × frequency
0	8	0 × 8 = ...
1	6	1 × 6 = ...
2	3	.... × .... = ...
3	1	.... × .... = ...
4	2	.... × .... = ...

Use the 'Number of goals × frequency' column in the table to work out the total number of goals. If this column is not given, add it on yourself.

(a) Write down the mode.

Guided The mode is .... goals.

The mode is the number of goals scored most often.

(1 mark)

(b) Find the median.

Guided The median will be the $\frac{20+1}{2}$ = ... th value.

The median is .... goal.

(1 mark)

(c) Work out the range.

Guided Range = largest value − smallest value

= 4 − ....

= .... goals

(2 marks)

(d) Work out the mean.

Guided Mean = $\frac{\text{total number of goals}}{20}$ = $\frac{\text{...........}}{20}$

= .... goals

Add up the final column to work out the total number of goals.

(3 marks)

D **2** Sarah had 10 boxes of sweets.
She counted the number of sweets in each box.
The table gives information about her results.

Number of sweets	Frequency
15	2
16	3
17	4
18	1

The 'modal number' is the same as the 'mode'.

(a) Write down the modal number of sweets in a box.

.... **(1 mark)**

(b) Find the median number of sweets in a box.

.... **(1 mark)**

(c) Work out the mean.

.... sweets **(3 marks)**

Averages from tables 2

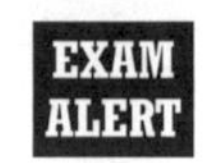

1 The table shows information about the numbers of minutes 40 children take to get to school.

Number of minutes (m)	Frequency	Midpoint	Midpoint × frequency
$0 < m \leqslant 10$	13	5	13 × 5 = ...
$10 < m \leqslant 20$	16	15	16 × 15 = ...
$20 < m \leqslant 30$	8	....	.... × .... = ...
$30 < m \leqslant 40$	3	....	.... × .... = ...

Work out an estimate for the mean number of minutes.

Exam questions similar to this have proved especially tricky – be prepared! ResultsPlus

$$\text{Mean} = \frac{\text{total number of minutes}}{40} = \frac{\ldots\ldots\ldots\ldots}{40}$$

= .... minutes

Add up the final column to work out the total number of minutes.
40 = total of the 'Frequency' column.

(4 marks)

2 The table shows information about the weights, in kilograms, of some children.

Weight (w kg)	Frequency		
$60 < w \leqslant 65$	6		
$65 < w \leqslant 70$	7		
$70 < w \leqslant 75$	10		
$75 < w \leqslant 80$	2		

(a) Write down the modal class.

.... **(1 mark)**

(b) Find the class interval that contains the median.

Add up the frequency column to work out the total number of children.

.... **(1 mark)**

(c) Work out an estimate for the mean weight.

.... kg **(4 marks)**

3 The table shows information about the total amounts, in pounds, some people spent in a supermarket.

Total amount (£T)	Frequency
$0 < T \leqslant 20$	7
$20 < T \leqslant 40$	10
$40 < T \leqslant 60$	13
$60 < T \leqslant 80$	5
$80 < T \leqslant 100$	5

Work out an estimate for the mean amount spent.

£..................... **(4 marks)**

Scatter graphs

1 Some students took both an English test and a science test.
The scatter graph shows information about their results.

(a) What type of correlation does this scatter graph show?

Guided

The scatter graph shows .... correlation.

(1 mark)

(b) Draw a line of best fit on the scatter graph.

(1 mark)

(c) Use your line of best fit to estimate:

(i) the science mark for a student with an English mark of 25

> Draw a vertical line from a mark of 25 for English to your line of best fit. Then draw a horizontal line across to find the science mark.

.... **(2 marks)**

(ii) the English mark for a student with a science mark of 27

.... **(2 marks)**

2 The scatter graph shows some information about the ages and values of ten cars.
The cars are the same make and type.

(a) What type of correlation does this scatter graph show?

....

(1 mark)

(b) Another car of the same make and type is 2 years old.
Estimate the value of this car.

£....................................

(2 marks)

> Draw a line of best fit on the graph to help you estimate the value or age of a car.

(c) Another car of the same make and type has a value of £2800
Estimate the age of this car.

.... years **(2 marks)**

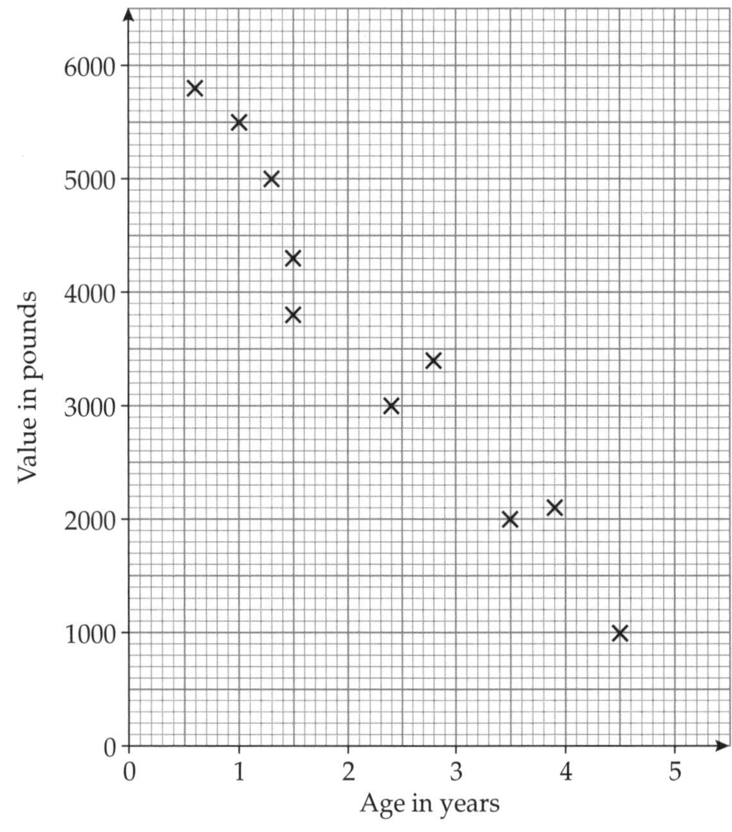

Probability 1

F **1** There are four counters in a bag.
Two counters are red, one counter is blue and one counter is green.

Impossible	Unlikely	Even	Likely	Certain

Which word from the box best describes the likelihood of each of these events?

(a) A counter which is taken at random is red.

.... **(1 mark)**

(b) A counter which is taken at random is yellow.

.... **(1 mark)**

(c) A counter which is taken at random is green.

.... **(1 mark)**

2 On the probability scale below, mark
(a) with the letter D, the probability that you will get a number less than 7 when you throw an ordinary dice
(b) with the letter T, the probability that when a fair coin is thrown once it comes down tails
(c) with the letter E, the probability that you will get an 8 when you throw an ordinary dice.

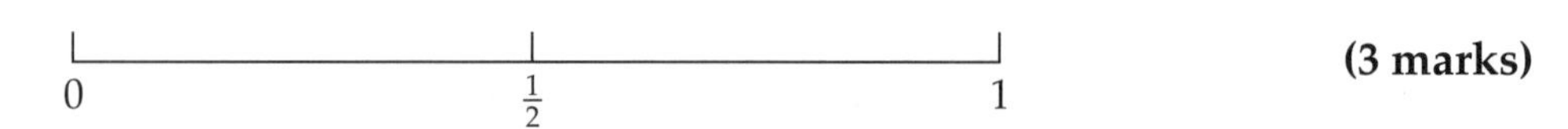

(3 marks)

3 Bella has a bag of 16 counters.
5 of the counters are blue. 3 of the counters are red. 8 of the counters are yellow.
Bella takes a counter at random from the bag.
Write down the probability that Bella

(a) takes a red counter

Probability Bella takes a red counter = $\frac{\dots\dots}{16}$ **(1 mark)**

(b) takes a blue counter

Probability Bella takes a blue counter = $\frac{\dots\dots}{16}$ **(1 mark)**

(c) takes a green counter.

.... **(1 mark)**

4 Rob has some cards. Each card has a letter on it.
The table shows the number of cards for each letter.

Letter	A	B	C	D
Number of cards	5	2	4	3

Rob takes a card at random.

(a) Write down the probability that he takes a card with the letter C.

Add up the number of each letter to find the total number of cards.

.... **(2 marks)**

(b) Write down the probability that he takes a card with the letter A.

.... **(2 marks)**

Probability 2

E 1 Manu eats at a café. He chooses one sandwich and one drink.
Write down all the possible combinations that Manu can choose.

Café menu	
Sandwiches	**Drinks**
Cheese	Tea
Egg	Juice
Ham	Water

Guided (C, T) (C, J) ....

....

....

Don't write out the full word each time – just use the first letter of each word.

(2 marks)

E 2 Jan throws an ordinary dice. She then throws a coin.
Write down all the combinations that Jan could get.

....

....

(2 marks)

E 3 A company makes torches. A torch is chosen at random. The probability that it has a fault is 0.04. Work out the probability that a torch, chosen at random, does **not** have a fault.

Guided Probability that the torch does **not** have a fault = 1 − ....

= ....

(1 mark)

D 4 Four teams, City, Rovers, Town and United play a competition to win a cup.
Only one team can win the cup.
The table below shows the probabilities of City or Rovers or Town winning the cup.

City	Rovers	Town	United
0.38	0.27	0.15	x

Work out the value of x.

Guided Probability(United) = 1 − Probability(not United)

= 1 − (0.38 + 0.27 + 0.15)

= 1 − ....

= ....

(2 marks)

D 5 A bag contains counters which are red or green or yellow or blue.
The table shows each of the probabilities that a counter taken at random from the bag will be red or green or blue.

Colour	Red	Green	Yellow	Blue
Probability	0.15	0.34		0.2

A counter is to be taken at random from the bag.

(a) Work out the probability that the counter will be red or green.

.... **(1 mark)**

(b) Work out the probability that the counter will be yellow.

.... **(2 marks)**

Probability 3

D 1 The probability that a biased dice will land on 2 is 0.3
Sue is going to roll the dice 60 times.
Work out an estimate for the number of times the dice will land on 2.

Guided

Estimate for number of 2s = ... × 60

= ....

(2 marks)

D 2 Andy counted the number of drawing pins in each of 20 boxes.
The table shows his results.

Number of drawing pins	28	29	30	31	32
Frequency	3	1	11	4	1

One of these boxes is taken at random.

(a) Write down the probability that the box contains exactly 31 drawing pins.

.... **(2 marks)**

Andy has 200 of these boxes of drawing pins.

(b) Work out an estimate for the number of these boxes that contain exactly 31 drawing pins.

.... **(2 marks)**

D 3 60 children went on a school trip.
They went to London or to New York.
26 boys and 15 girls went to London.
12 boys went to New York.

(a) Use this information to complete the two-way table.

Guided

	London	**New York**	**Total**
Boys	26		
Girls	15		
Total			60

(3 marks)

One of these 60 children is chosen at random.

(b) Write down the probability that this child went to New York.

.... **(2 marks)**

C 4 Marco has a 4-sided spinner. The sides of the spinner are numbered 1, 2, 3 and 4
The spinner is biased.
The table shows the probabilities that the spinner will land on each of the numbers 1, 2 and 4

Number	1	2	3	4
Probability	0.2	0.35		0.18

First work out the probability of getting a 3.

Marco spins the spinner 400 times.
Work out an estimate for the number of times he gets a 3

.... **(3 marks)**

Problem-solving practice

1 The table shows information about the numbers of boys and girls choosing each of four activities.

	Canoeing	Archery	Climbing	Sailing
Boys	5	8	6	4
Girls	9	7	4	7

On the grid, draw a suitable chart or diagram to display this information.

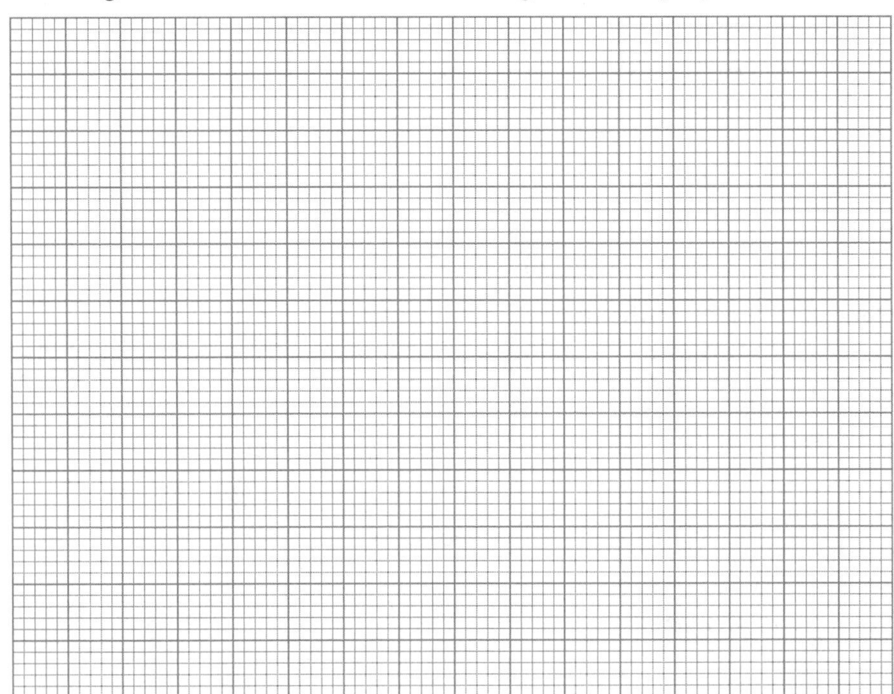

One suitable chart would be a dual bar chart. Remember to label the axes and use a key.

(4 marks)

2 Grace rolls an ordinary dice. She then flips a fair coin.
List all the possible outcomes she could get.

It is sensible to use abbreviations; for example, use H for 'head' and T for 'tail'.

Guided (1, H), (1, T) ....

....

(2 marks)

3 James asked some people in a café to name their favourite drink.
The pie chart shows his results.

8 people named squash as their favourite drink.
How many people named cola as their favourite drink?

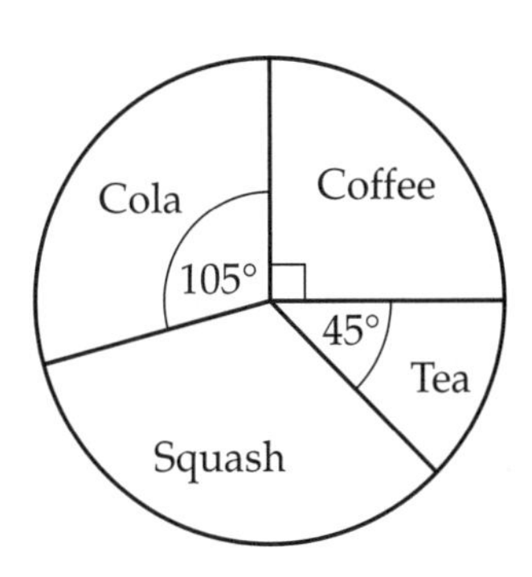

Start by working out the size of the angle for squash.

.... **(3 marks)**

Problem-solving practice

D 4 | ? | ? | ? | ? | ? |

Kunal has five cards. Each card has a number on it. The numbers are hidden.
The mode of the five numbers is 2
The median of the five numbers is 5
The total of the five numbers is 22
Work out the numbers on the cards.

The median is 5 so put 5 on the middle card. The mode is 2 so this must be the number that occurs most often **but** there can only be two numbers that are less than 5. The last two numbers must be bigger than 5

.... **(3 marks)**

D 5 The scatter graph shows some information about eight cars.
For each car it shows the engine size, in litres, and the distance, in miles, it travels on one gallon of petrol.

A car has an engine size of 3.5 litres.
Estimate how far this car will travel on one gallon of petrol.

.... **(2 marks)**

Draw a line of best fit on the scatter graph.

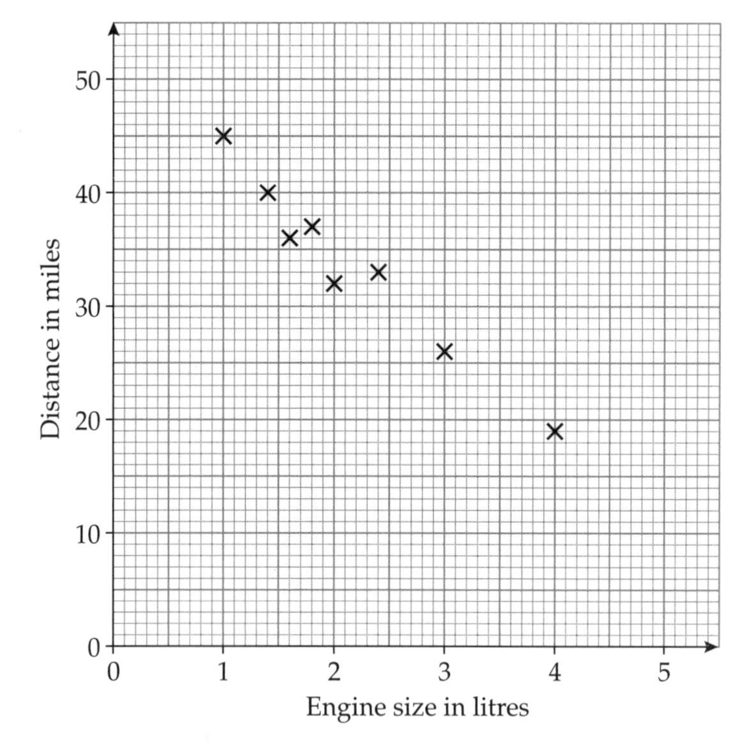

C 6 All students in class 9A study either French or German or Spanish.

21 of the 40 students are boys.
7 of the boys study Spanish.
9 girls study French.
8 of the 10 students who study German are boys.

Work out the number of students who study Spanish.

Guided

	French	German	Spanish	Total
Boys				
Girls				
Total				40

.... students study Spanish. **(4 marks)**

Formulae page

Area of trapezium $= \frac{1}{2}(a + b)h$

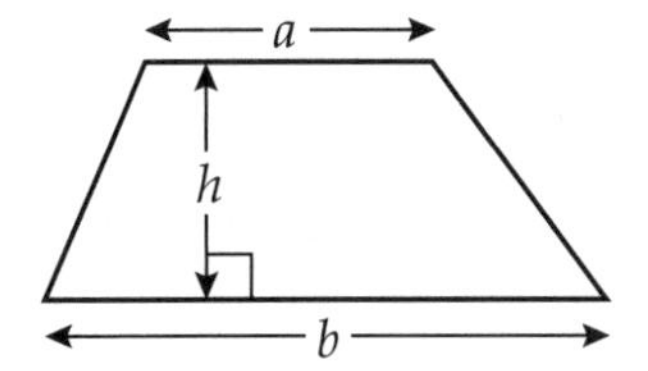

Volume of a prism = area of cross section × length

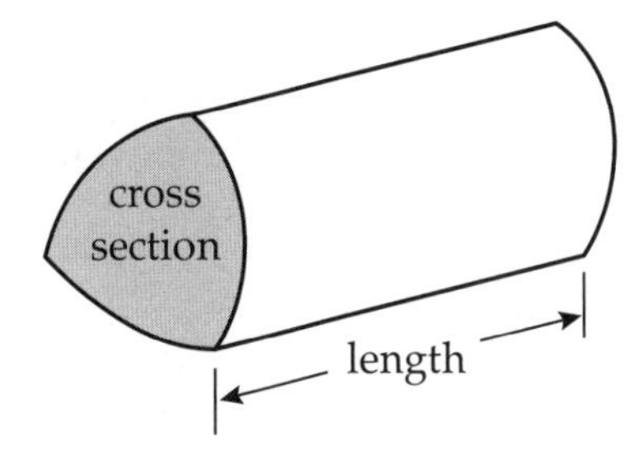

Paper 1

Practice exam paper

Foundation Tier
Time: 1 hour 45 minutes
Calculators must not be used

1 Kunal asked some students to tell him their favourite zoo animal.
He used the information to draw this bar chart.

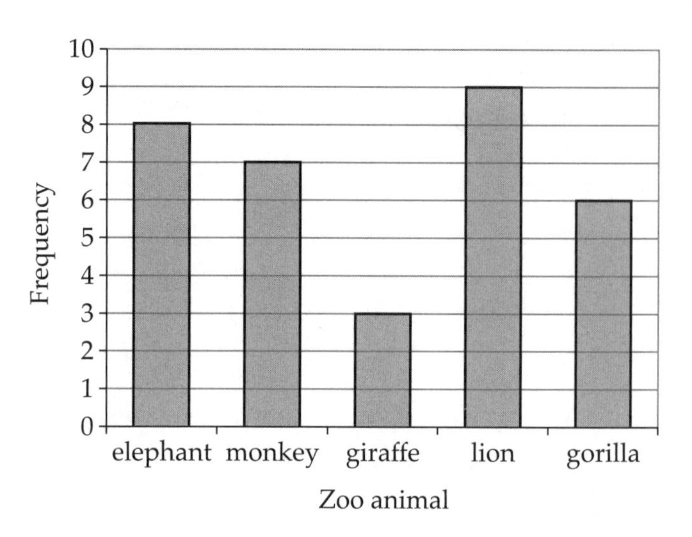

(a) How many students said monkey?

.... **(1 mark)**

(b) Which zoo animal did most students say?

.... **(1 mark)**

(c) Work out the number of students that Kunal asked.

.... **(2 marks)**

2 (a) Write the number **five thousand and twenty six** in figures.

.... **(1 mark)**

(b) Write down the value of the 8 in the number 45 983

.... **(1 mark)**

(c) Write the number 3278 correct to the nearest thousand.

.... **(1 mark)**

(d) Write these numbers in order of size. Start with the smallest number.

0.7 0.09 0.79 0.0971

.... **(1 mark)**

3 The table shows some information about five cars.

Make	Age (years)	Colour	Number of doors
Ford	3	red	4
Rover	2	blue	5
Volvo	2	silver	4
Golf	5	silver	3
Mazda	4	black	4

(a) Write down the make of the car with 5 doors.

.... **(1 mark)**

(b) Which make of car has 4 doors and is silver?

.... **(1 mark)**

Alan wants to buy a car that has 4 doors and is **less** than 4 years old.

(c) Write down the makes of the cars that he could buy.

.... **(1 mark)**

4 Here are some quadrilaterals.

A B C D E F

(a) Write down the mathematical name for shape **C**.

.... **(1 mark)**

(b) Write the letter of the shape that is a rhombus.

.... **(1 mark)**

(c) Write down the number of lines of symmetry of shape **A**.

.... **(1 mark)**

(d) Write down the order of rotational symmetry of shape **B**.

.... **(1 mark)**

***5** There are 38 people on a coach.
15 people got off at the first stop. 13 people got on at the first stop.
11 people got off at the second stop. 19 people want to get on at the second stop.
The coach holds a maximum of 42 people.
Can all the 19 people get on the coach? You must show your working.

(4 marks)

6 Here are some patterns made from sticks.

Pattern number 1 Pattern number 2 Pattern number 3

(a) Draw Pattern number 4 in the space below.

(1 mark)

(b) How many sticks are needed for Pattern number 8?

.... **(2 marks)**

Harry says that he will need 80 sticks for Pattern number 20.

(c) Is Harry correct? You must give a reason for your answer.

....

....

(2 marks)

7 Ibrahim writes down three prime numbers. All his prime numbers are different.
Ibrahim adds up his three prime numbers. The total is an odd number between 25 and 30
What three prime numbers could Ibrahim have written down?

.... **(3 marks)**

8 Sarah counted the number of birds in her garden at 7 am on each of 20 days.
Here are her results.

3	4	6	5	3	2	2	4	3	4
2	5	1	3	4	4	6	2	4	1

(a) Complete the frequency table.

Number of birds	Tally	Frequency
1		
2		
3		
4		
5		
6		

(2 marks)

(b) Write down the mode.

.... **(1 mark)**

9 Here is a café menu.
Rick is going to choose one drink and one sandwich.

Café menu

Drinks		**Sandwiches**	
Tea	£1.20	Ham	£2.45
Coffee	£1.80	Egg	£2.30
Water	£1.10	Prawn	£2.80

(a) Write down all the possible combinations Rick could choose.

....

....

(2 marks)

Lucy buys
1 cup of tea
1 cup of coffee
2 egg sandwiches
She pays with a £10 note.

(b) Work out how much change Lucy should get.

£........................ **(3 marks)**

10 (a) Complete this table.

Write a sensible unit for each measurement. Three have been done for you.

	Metric	**Imperial**
The amount of petrol in a tank	litres	....
The distance between France and England	....	miles
The length of your foot	centimetres	....

(3 marks)

(a) Change 5 kg into grams.

.... grams **(1 mark)**

(b) Change 450 cm into metres.

.... m **(1 mark)**

11 The table shows the midday temperatures in four different cities on Saturday.

City	London	York	Cardiff	Glasgow
Midday temperature (°C)	2	–3	–5	–2

(a) Which city had the lowest temperature?

.... **(1 mark)**

(b) Work out the difference between the temperature in York and the temperature in London.

.....................°C **(1 mark)**

By Sunday the midday temperature in Cardiff had risen by 8 °C.

(c) Work out the midday temperature in Cardiff on Sunday.

.....................°C **(1 mark)**

12 (a) A shaded shape is drawn on a square grid.

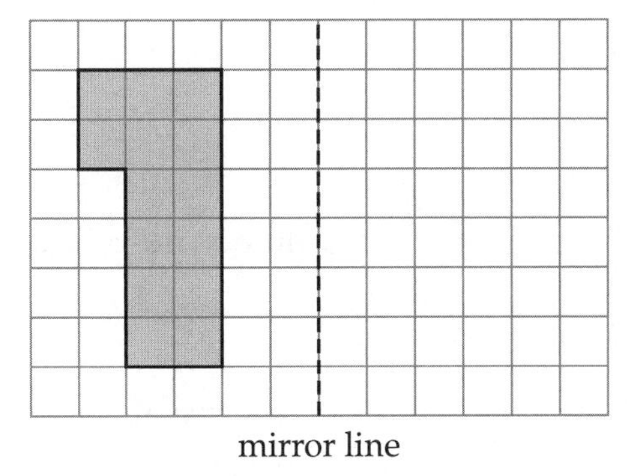

Reflect the shaded shape in the mirror line. **(2 marks)**

(b)

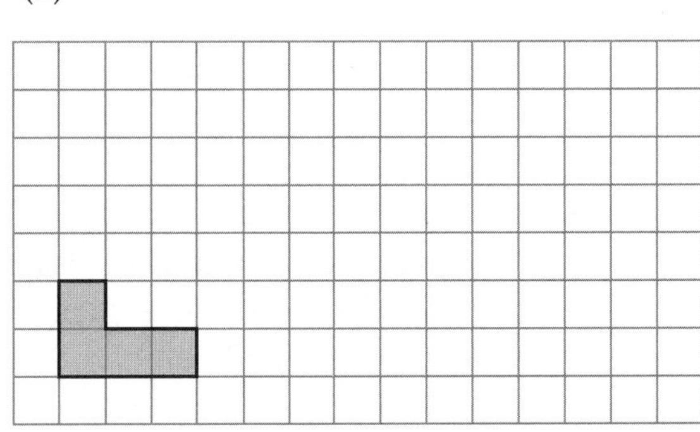

Enlarge the shaded shape with a scale factor of 3 **(2 marks)**

***13** ABC is an isosceles triangle.

ACD is a straight line.

Work out the size of the angle marked x.

Give reasons for your answer.

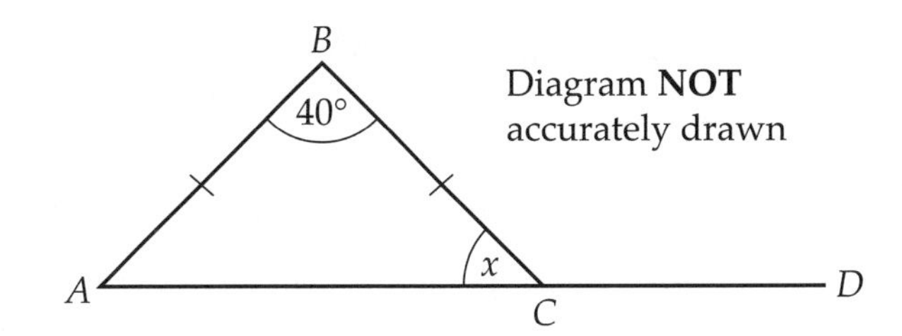

(4 marks)

14 (a) Work out $\frac{3}{7} \times \frac{2}{5}$

.... **(1 mark)**

(b) Work out $\frac{7}{12} - \frac{1}{4}$

.... **(2 marks)**

15 Here is a cuboid.
Work out the volume of the cuboid.

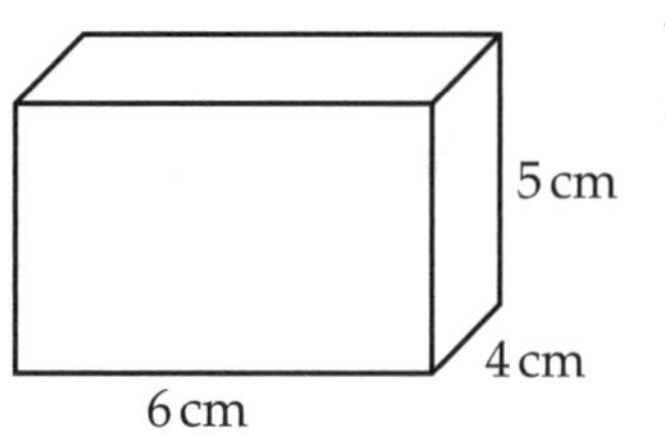

Diagram **NOT** accurately drawn

.... **(3 marks)**

16 $P = 2l + 2w$
Work out the value of P when $l = 6$ and $w = 10$

$P =$ .... **(2 marks)**

17 The diagram shows a sketch of triangle ABC.
$AB = 7$ cm. $BC = 5.3$ cm. Angle $B = 42°$.
Make an accurate drawing of triangle ABC.
The line AB has been accurately drawn.

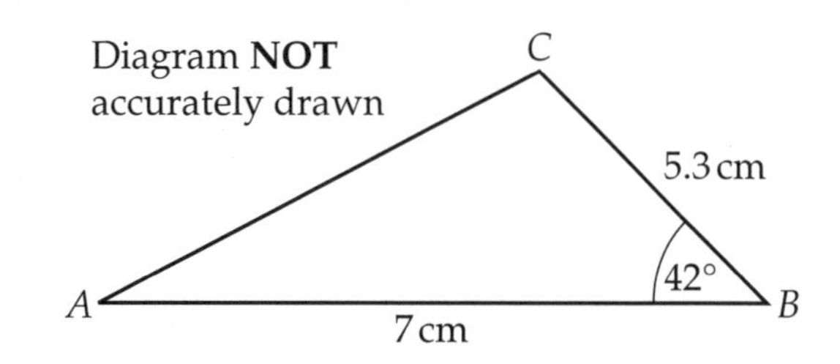

A ———————————————— B

(2 marks)

18 There are 12 counters in a bag.
5 of the counters are red. 4 of the counters are blue. 3 of the counters are yellow.
Fred takes a counter at random from the bag.
Write down the probability that the counter is

(a) red

....

(b) red or yellow

....

(c) green.

....

(3 marks)

19 (a) Solve $x + 6 = 13$

$x =$ .... **(1 mark)**

(b) Solve $\frac{y}{2} = 18$

$y =$ .... **(1 mark)**

(c) Solve $6m - 5 = 19$

$m =$ .... **(2 marks)**

(d) Solve $5w + 3 = 3w + 10$

$w =$ .... **(2 marks)**

20 Mrs Jones is organising a school trip. Here are her costs:

28 train tickets	£25 **each**
Coach hire	£240
Entrance fees	£175

The school will pay £500 towards the cost of the trip.
The rest of the cost will be shared equally amongst the students going on the trip.
There are 25 students going on the trip.
How much will each student have to pay?

£.................... **(5 marks)**

21 Work out an estimate for the value of $\frac{597.8}{1.92 \times 3.44}$

.... **(2 marks)**

22 On the grid, draw the graph of $y = 3x - 4$ for values of x from -1 to 3

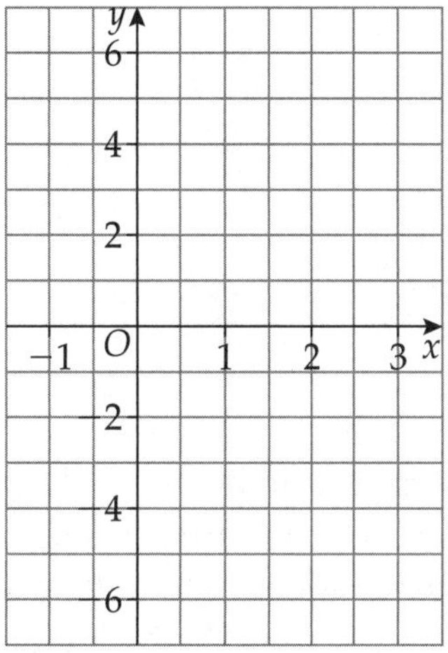

(3 marks)

23 (a) Berwyn is going to carry out a survey about magazines.
He is going to use this question on a questionnaire.

How much do you spend on magazines?

☐ £2–£5 ☐ £5– £10 ☐ more than £10

Write down **two** things that are wrong with this question.

1 ....

2 ....

(2 marks)

Berwyn wants to find out how many magazines people buy.

(b) Design a suitable question Berwyn could use for his questionnaire.

(2 marks)

24 The table gives information about the costs of sending a letter by first-class post and by second-class post.

First class	**Second class**
46p	36p

In one week, a company sent 400 letters.
The ratio of the number of letters sent first class to the number of letters sent second class was 5 : 3
Work out the total cost of sending all the letters.

£....................... **(4 marks)**

25 The diagram shows two towns, Alton and Beescroft.

× B

A ×

Scale: 1 cm represents 10 km

A new airport is to be built. The airport must be
less than 70 km from Beescroft **and**
closer to Alton than to Beescroft.
On the diagram, shade the region where the airport could be built. **(3 marks)**

26 Mr Smith bought a television. The usual price of the television was £850
The television was in a sale. In the sale, all prices were reduced by 20%.
Mr Smith paid £250 when he got the television.
He paid the rest of the cost of the television in 10 equal monthly payments.
Work out the amount of each monthly payment.

£.................... **(5 marks)**

27 140 children went on an activity day.
Each child chose one activity from climbing, canoeing or archery.
58 of the children were girls. 23 of the boys did climbing. 12 girls did archery.
17 of the 38 children who did canoeing were girls.
Work out the number of children who did climbing.

.... **(4 marks)**

Paper 2

Foundation Tier
Time: 1 hour 45 minutes
Calculators may be used

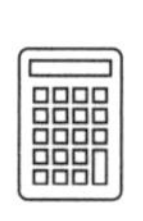

Practice exam paper

1 The pictogram shows the numbers of books sold by a shop on Monday, Tuesday and Wednesday one week.

Monday	
Tuesday	
Wednesday	
Thursday	
Friday	

Key: represents 8 books

(a) Write down the number of books sold on Monday.

.... **(1 mark)**

(b) Write down the number of books sold on Wednesday.

.... **(1 mark)**

16 books were sold on Thursday. 30 books were sold on Friday.

(c) Use this information to complete the pictogram. **(2 marks)**

2 (a) Write down the coordinates

(i) of the point A

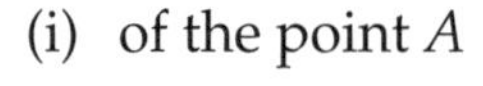

(.........., ...) **(1 mark)**

(ii) of the point B

(.........., ...) **(1 mark)**

(b) On the grid, plot the point $(-2, 3)$.
Label the point D. **(1 mark)**

(c) Find the coordinates of the midpoint of the line AC.

(.........., ...) **(2 marks)**

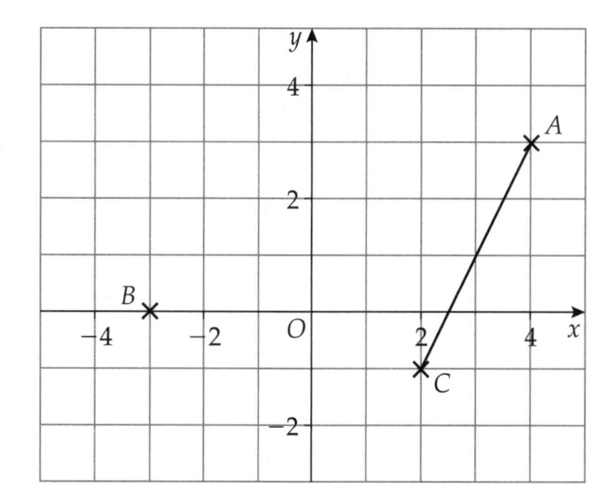

3 (a) Simplify $c + c + c + c + c$

.... **(1 mark)**

(b) Simplify $5x + 3y - 2x + 7y$

.... **(2 marks)**

***4** Mrs Smith is going to take her 4 children to the cinema.
Mrs Smith has a voucher to use at the cinema.
She can use her voucher to pay for cinema tickets.
The voucher is worth £30
Will the voucher be enough to pay for all the tickets?
You must show all your working.

Cinema Tickets	
Adult ticket	£8.65
Child ticket	£5.40

(3 marks)

5 (a) Measure the length of AB. A ——————————— B

.... **(2 marks)**

(b)

(i) What type of angle is shown by the letter c?

.... **(1 mark)**

(ii) Measure the size of angle c.

....................° **(1 mark)**

6 The table shows some information about the numbers of boys and girls taking part in four different sports.

	Swimming	**Tennis**	**Athletics**	**Volleyball**
Boys	9	6	14	13
Girls	12	11	10	5

The PE department want to display this information.
On the grid, draw a suitable diagram or chart.

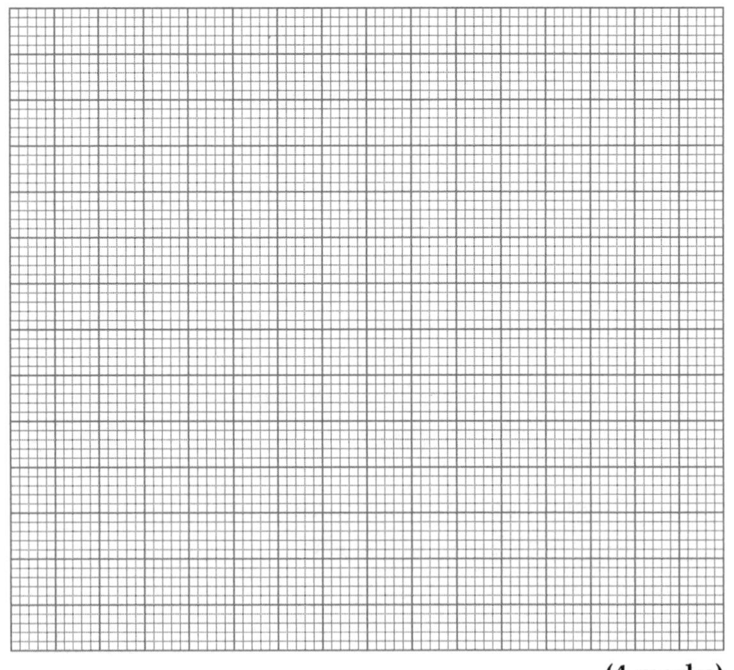

(4 marks)

7 600 students are going on a school trip by bus. Each bus can carry 35 students.
Work out the smallest number of buses needed to carry all the students.

.... **(3 marks)**

8 Six shapes are drawn on a square grid.
Two of the shapes are congruent.

(a) Write down the letters of these two shapes.

.... and .... **(1 mark)**

One of the shapes is an enlargement of shape **A**.

(b) Write down the letter of this shape.

.... **(1 mark)**

Here is a rectangle.

(c) Draw in all the lines of symmetry on the rectangle.

(2 marks)

9 Kerry picks flowers. She puts the flowers in bunches.
She packs the bunches of flowers into boxes. There are 8 bunches of flowers in a box.
On one weekend, Kerry fills 34 boxes with bunches of flowers on Saturday and she fills 45 boxes with bunches of flowers on Sunday.
Kerry gets paid 9p for each bunch of flowers she picks.
How much money, in total, did Kerry get paid for her work this weekend?

£.................... **(4 marks)**

10 Lesley rolled a six-sided dice 10 times. Here are her results.

1 2 2 2 3 3 4 5 5 6

(a) Work out the range.

.... **(2 marks)**

(b) Work out the median.

.... **(1 mark)**

(c) Work out the mean.

.... **(2 marks)**

11 (a) Write $\frac{3}{4}$ as a decimal.

.... **(1 mark)**

(b) Write 0.25 as a fraction.

.... **(1 mark)**

(c) Write 90 as a fraction of 105. Give your fraction in its simplest form.

.... **(2 marks)**

12 Use your calculator to work out $\sqrt{384} + 6.7^2$
Write down all the figures on your calculator display.

.... **(2 marks)**

13 The diagrams show some solid shapes and their nets.
An arrow has been drawn from one solid shape to its net.
Draw an arrow from each of the other solid shapes to its net.

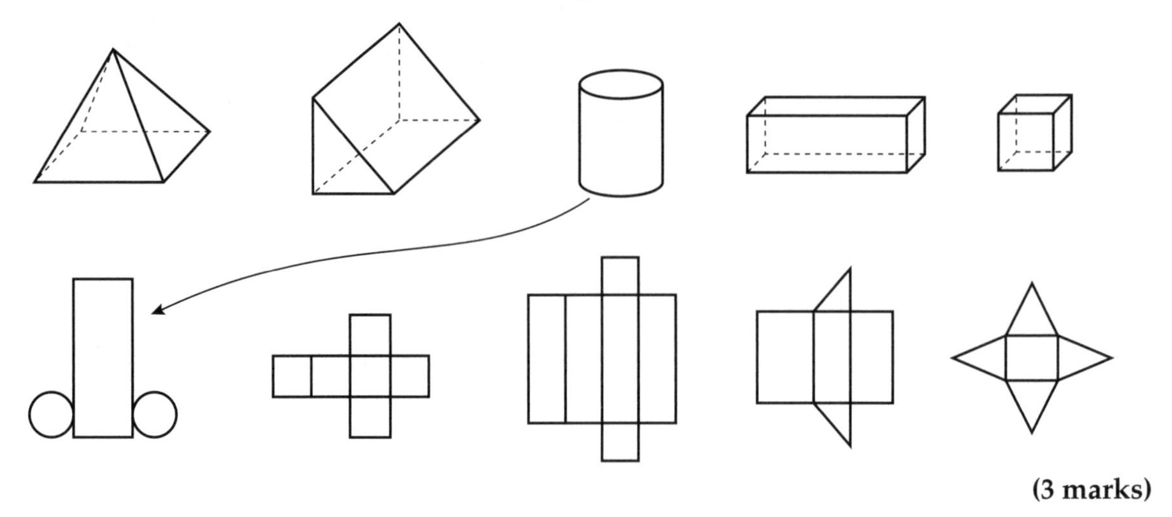

(3 marks)

14 The table gives information about the numbers of votes four people received in an election.

Name	Number of votes
Jean	144
Julie	180
Harry	90
Gwen	126

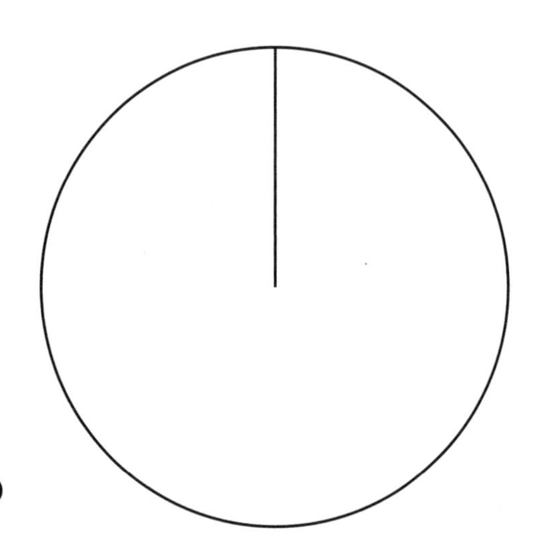

Draw an accurate pie chart to show these results.

(3 marks)

15 Linda has a monthly pay plan for her mobile phone.
Each month, she pays a total of £12.95 plus the cost of any extra minutes and any extra texts.
Last month, Linda used 154 minutes and sent 1021 texts.
Work out how much she paid in total last month.

Monthly Pay Plan
For **£12.95** per month you get: 120 minutes and 1000 texts
Extra minutes: 17.5p for each minute
Extra texts: 5p for each text

£.................... **(5 marks)**

16 Pavinder wants to carry out a survey to find out which football team people support.
Design a suitable data collection table he could use to collect this information.

(3 marks)

17 You can use this graph to change between metres and feet.

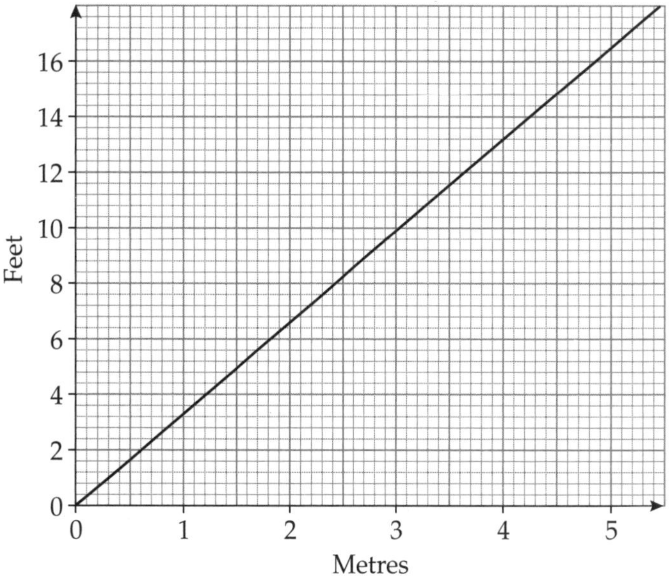

(a) Use the graph to change 3 metres into feet.

.... feet **(1 mark)**

(b) Use the graph to change 13 feet into metres.

.... metres **(1 mark)**

Kim throws a ball 60 feet.
Liz throws a ball 18 metres.

*(c) Who throws the ball the greatest distance? You must explain your answer.

(3 marks)

18 The manager of a shop uses this rule to work out the cost of photocopying.

Cost of photocopying in pence = number of copies × 15 + 65

Alex wants to make 40 copies.

(a) Work out the cost of Alex's photocopying.

.....................p **(2 marks)**

Ben does some photocopying. His total cost is £4.55

(b) Work out the number of copies Ben got.

.... **(3 marks)**

19 There are 300 sweets in a bag. The sweets are red or orange or yellow.
$\frac{1}{6}$ of the sweets are red. $\frac{3}{5}$ of the sweets are orange.
Work out the number of yellow sweets in the bag.

.... **(4 marks)**

20 Here is a list of ingredients needed to make 8 pancakes.
Sally wants to make 20 pancakes.
Work out the amount of each ingredient she will need.

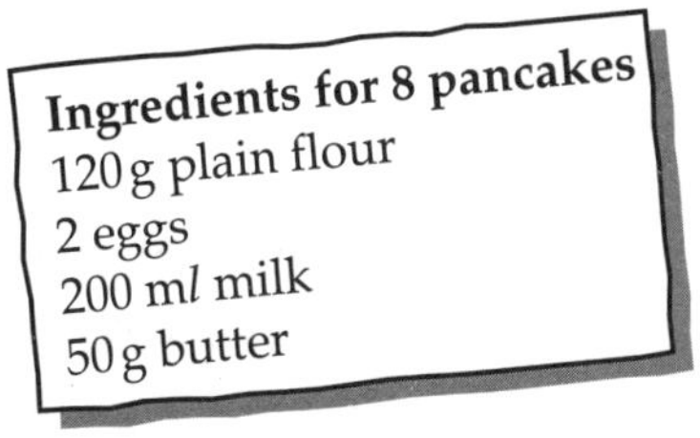

Ingredients for 8 pancakes
120 g plain flour
2 eggs
200 m*l* milk
50 g butter

Plain flour .... g Eggs ....

Milk .... m*l* Butter .... g **(3 marks)**

21 The cost of 5 kg of apples is £4.60
The cost of 3 kg of apples and 4 kg of pears is £6.16
Work out the cost of 1 kg of pears.

.... **(4 marks)**

22 The diagram shows a shape.
Work out the area of the shape.

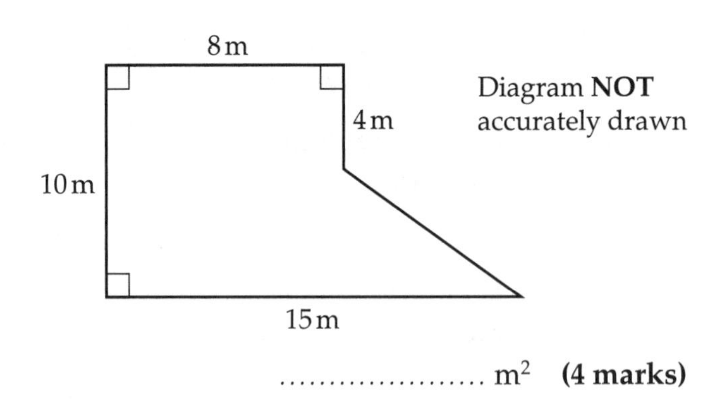

.... m^2 **(4 marks)**

23 Barry sells DVDs. He sells each DVD for £8.50 plus VAT at 20%. He sells 1300 DVDs.
Barry gets a bonus. His bonus is 4% of his total sales. Work out how much bonus Barry gets.

£.................... **(5 marks)**

24 Describe fully the single transformation that maps shape **P** onto shape **Q**.

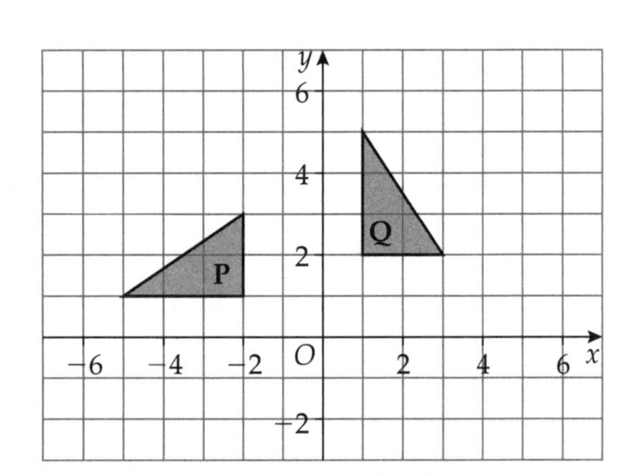

....

....

....

(3 marks)

***25** The diagram shows the floor of a room.
The floor is in the shape of a semicircle of diameter 16 m.
The floor is to be varnished.
John is going to put two coats of varnish on the floor.
One tin of varnish will cover an area of 20 m^2.
John has bought 10 tins of varnish. Will John have enough varnish?

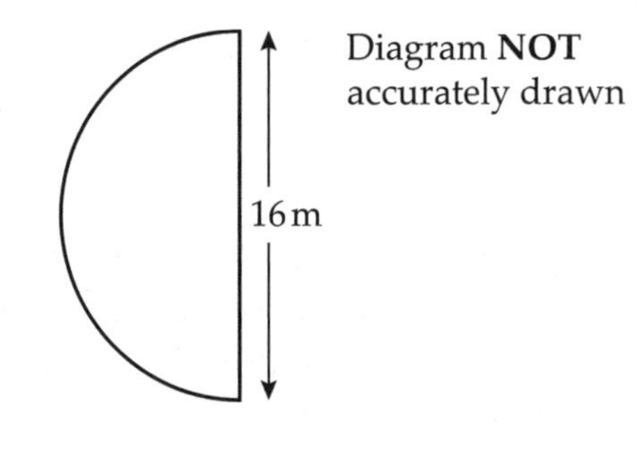

(4 marks)

26 The equation $x^3 - 6x = 140$ has a solution between 5 and 6
Use a trial and improvement method to find this solution.
Give your answer correct to 1 decimal place.
You must show **ALL** your working.

$x =$ .... **(4 marks)**

Answers

The number given to each topic refers to its page number.

NUMBER

1. Place value

1. (a) seven thousand and eighty-two
   (b) 4700 (c) 5 tens or 50 (d) 3117
2. (a) 43 000 (b) 3 thousands or 3000
3. (a) 78, 84, 470, 478, 784
   (b) 1435, 1453, 3451, 4531, 5431
4. (a) 4 000 000
   (b) (i) three hundred and fifty thousand
   (ii) 8 hundreds or 800
   (c) (i) 39 000
   (ii) 38 900
5. 78 900, 79 000, 98 000, 780 000, 781 900

2. Rounding numbers

1. (a) 5000 (b) 23 400 (c) 6730
2. (a) 50 000 (b) 0.06
3. (a) 3 (b) 70 000 (c) 0.090
4. (a) 7 (b) 7.0 (c) 7.03
5. (a) 0.009 (b) 0.0093 (c) 0.009 28

3. Adding and subtracting

1. (a) 796 (b) 474
2. (a) 6925 (b) 3466
3. (a) 1472 (b) 143
4. £86 5. 34 people 6. 27 people

4. Multiplying and dividing

1. (a) 1380 (b) 171
2. £432
3. 884
4. (a) 19 740 (b) 25
5. £23

5. Decimals and place value

1. (a) 6 tenths (b) 9 hundredths
2. 1.9, 3.5, 5.6, 7, 8.3, 8.7
3. 0.04, 0.045, 0.45, 0.5, 0.541
4. 0.8, 0.815, 0.82, 0.879, 0.89
5. (a) 1972 (b) 19.72
6. (a) 332.8 (b) 33.28 (c) 1280

6. Operations on decimals

1. (a) 33.56 (b) 71.72
2. (a) 145.24 (b) 71.66
3. (a) 0.18 (b) 24.4
4. £19.64 5. £192.92 6. 13.364

7. Estimating answers

1. 60 2. 30 3. 100
4. 120 5. 160 6. 2000

8. Negative numbers

1. (a) −6, −5, −2, 0, 4 (b) (i) −6 (ii) 15
2. (a) −9, −8, −5, −2, 2, 3 (b) (i) 18 (ii) −5
3. (a) 5 °C (b) 9 °C (c) −6 °C
4. −1 °C is half way so Sasha is wrong

9. Squares, cubes and roots

1. (a) 36 (b) 8 (c) 10
2. (a) 49 (b) 5 (c) 64 (d) 2
3. (a) 64 (b) 125 (c) 9 (d) 4
4. (a) 9000 (b) 7.1
5. counter example, e.g. 9 and 25 are both square numbers but $9 + 25 = 34$ which is even
6. Jerry is right because $4^3 = 4 \times 4 \times 4 = 64$; $8^2 = 8 \times 8 = 64$

10. Factors, multiples and primes

1. 1, 2, 3, 4, 6, 8, 12, 24
2. (a) 4 (b) 25 or 15 (c) 7 (d) 4 and 18
3. (a) 1, 2 and 4 (b) 1, 2, 4 and 8 (c) 1 and 5
4. $2 \times 2 \times 3 \times 5$ or $2^2 \times 3 \times 5$
5. $2 \times 3 \times 5 \times 5$ or $2 \times 3 \times 5^2$
6. $2 \times 2 \times 2 \times 3 \times 3$ or $2^3 \times 3^2$

11. HCF and LCM

1. 60
2. (a) (i) $2 \times 3 \times 3 \times 5$ (ii) $2 \times 2 \times 2 \times 3 \times 5$
   (b) $2 \times 3 \times 5 = 30$
3. (a) $2 \times 2 \times 3 \times 7$ or $2^2 \times 3 \times 7$
   (b) 12
4. 240

12. Fractions

1. (a) $\frac{2}{5}$ (b) £13
2. (a) $\frac{1}{3}$ (b) $\frac{4}{8}, \frac{7}{14}$
3. 18 kg 4. £40 5. £210

13. Simple fractions

1. (a) $\frac{11}{20}$ (b) $\frac{7}{36}$
2. (a) $\frac{5}{9}$ (b) $\frac{10}{11}$
3. (a) $\frac{11}{15}$ (b) $\frac{8}{15}$
4. $\frac{1}{12}$
5. 5

14. Mixed numbers

1. $7\frac{13}{20}$
2. (a) $11\frac{3}{14}$ (b) $3\frac{5}{8}$
3. $1\frac{5}{14}$
4. (a) $\frac{3}{4}$ (b) $2\frac{2}{9}$

15. Number and calculator skills

1. £13.70 2. £14.34 3. 14
4. 24 5. 48.36
6. 41.446 833 41 7. 1.451 013 451

16. Percentages

1. (a) $\frac{47}{100}$ (b) $\frac{3}{10}$
2. £128
3. (a) 15 m (b) £80
4. £100
5. (a) 62.5% (b) 9
6. 60 students

17. Percentage change

1. £54.60 2. £77.33 3. £504
4. £252 5. 35p

18. Fractions, decimals and percentages

1. $\frac{2}{5}, \frac{9}{20}$, 47%, $\frac{1}{2}$, 55%
2. $\frac{13}{20}, \frac{7}{10}$, 72%, $\frac{3}{4}$, 76%
3. £10 400
4. £40

19. Ratio

1. 3 : 5 2. 7 : 3
3. No, she has got the ratio the wrong way round. The ratio should be 5 : 3
4. £90 5. 720 km
6. 150 g jar – 1 g costs 1.9p; 200 g – 1 g costs 1.95p
   So 150 g jar is better value

20–21. Problem-solving practice

1. £2.45
2. 5 coaches
3. £12
4. 18:45
5. Cheap Tickets (£37.63, compared with £37.80 at Tickets R-US)
6. £10 400

ALGEBRA

22. Collecting like terms

1. (a) $3t$ (b) $3bc$ (c) $11k + 3m$
2. (a) $4m + 2p$ (b) $4x$ (c) $2p^2$
3. (a) $11a + 5b$ (b) $6e + 7f$
4. Sam is right because it contains an equals sign
5. $2x - 10y$
6. (a) $7 + 7a$ (b) $4p - 5t$
7. (a) $6xy$ (b) $2r + 8t + 4$

23. Simplifying expressions

1. (a) r^3 (b) $3mp$ (c) $20xy$
2. (a) a^5 (b) $8kn$ (c) $18hj$
3. (a) e^2f^3 (b) $32xy$
4. (a) $3a$ (b) $6y$
5. (a) $3b$ (b) $7p$
6. (a) $28gh$ (b) $8tz$ (c) $8m$ (d) $8abc$

24. Indices

1. (a) m^{12} (b) p^8 (c) t^{20}
2. (a) d^{13} (b) e^4 (c) f^8

3. (a) $14g^9$ (b) $4b^3$
4. (a) g^7 (b) k^8
5. (a) x^3 (b) y^8
6. (a) h^2 (b) a^6

25. Expanding brackets

1. (a) $4a - 8b$ (b) $d^2 + 5d$
2. (a) $15b + 3$ (b) $e^2 - 2e$
3. (a) $2d^2 - 6d$ (b) $3f - 3fe$
4. (a) $18 - 24d$ (b) $3p^3 - p^2$
5. $10x + 7y$
6. (a) $6k + 10$ (b) $5x - y$
7. $10x$

26. Factorising

1. (a) $3(x + 6)$ (b) $y(y - 9)$
2. (a) $5(m + 4)$ (b) $2(4t - 7)$ (c) $v(v - 1)$
3. (a) $3(3 - k)$ (b) $m(2m - 1)$
4. (a) $8(2p - 3)$ (b) $4x(x + 2)$
5. $4m(2p - 3m)$
6. (a) $7b(2a - 3c)$ (b) $2x(4x + 5y)$
7. (a) $4x(3y - 4)$ (b) $20y(1 - 2y)$

27. Sequences

1. (a) 19, 23 (b) Add 4 each time (c) 35
2. (a) 54, 48
   (b) He is wrong because the sequence goes 60, 54, 48, 42, 36
3. (a) drawing of pattern number 4
   (b) 17, 21
   (c) 37
   (d) No, because 100 is an even number; the number of sticks needed is always an odd number

28. *n*th term of a sequence

1. $4n - 3$
2. $-5n + 22$ or $22 - 5n$
3. (a) $4n - 2$
   (b) 73 is an odd number, all numbers in sequence are even numbers
4. (a) 11, 19, 27
   (b) 5th term is 43, 6th term is 51 therefore 45 is not a term
5. 149

29. Equations 1

1. (a) $k = 5$ (b) $m = 19$
2. (a) $p = 3$ (b) $x = 35$
3. (a) $x = 6$ (b) $y = 2.5$
4. (a) $h = 8$ (b) $m = 6$
5. $x = 6$
6. (a) $y = 4$ (b) $x = 2$
7. $x = 24$

30. Equations 2

1. (a) $x = 2$ (b) $x = 7$
2. (a) $x = 4$ (b) $x = 0.5$
3. $x = -\frac{1}{4}$
4. (a) $x = 3\frac{1}{4}$ (b) $x = -17$
5. $x = 3\frac{1}{5}$
6. $x = -3$

31. Writing equations

1. 140° 2. 25° 3. 7.5 cm

32. Trial and improvement

1. 2.4 2. 3.6 3. 4.2

33. Inequalities

1. $x > -1$ 2. $x \leq 2$
3. −2, −1, 0, 1, 2 4. −5, −4, −3, −2, −1, 0
5.

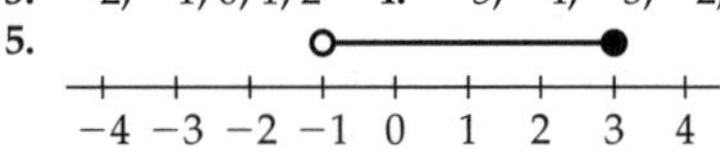

6.

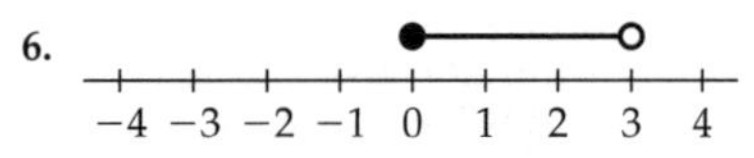

34. Solving inequalities

1. $x > 6$
2. (a) $x < 6$ (b) $x = 5$
3. (a) $x < 2$
   (b)

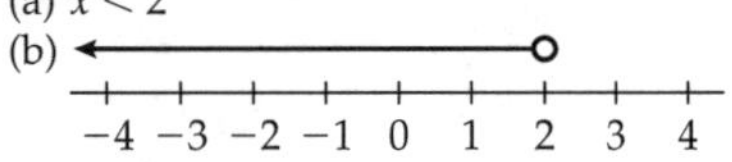

4. $x \geq 4\frac{2}{3}$
5. $x < 0.5$

35. Substitution

1. (a) 17 (b) 29
2. (a) 27 (b) 24
3. (a) 48 (b) 40
4. 192
5. (a) 69 (b) −12

36. Formulae

1. £66 2. 110 minutes 3. £205
4. 57 5. 84

37. Writing formulae

1. $T = 30b$
2. $E = 12x$
3. $C = 45 + 30t$
4. $C = 6c + 4b$
5. $P = 8x - 5$
6. $C = 0.01 \times (4a + 3b)$ or equivalent

38. Rearranging formulae

1. 9 days 2. $a = 14$ 3. $a = 6$
4. $t = \frac{v - u}{6}$ 5. $a = \frac{c + t^2}{2}$ 6. $k = \frac{p + 6}{2}$

39. Coordinates

1. (a) (i) (2, 3) (ii) (−3, −2)
   (b) points C and D plotted
2. (a) (i) (1, 2) (ii) (5, 6)
   (b) (3, 4)
3. (5, 5)
4. (5, 3)

40. Straight-line graphs 1

1. 2
2. −0.5
3. (a) 3 (b) −17

41. Straight-line graphs 2

1. (a)

x	−1	0	1	2	3
y	**−4**	−1	**2**	**5**	8

(b)

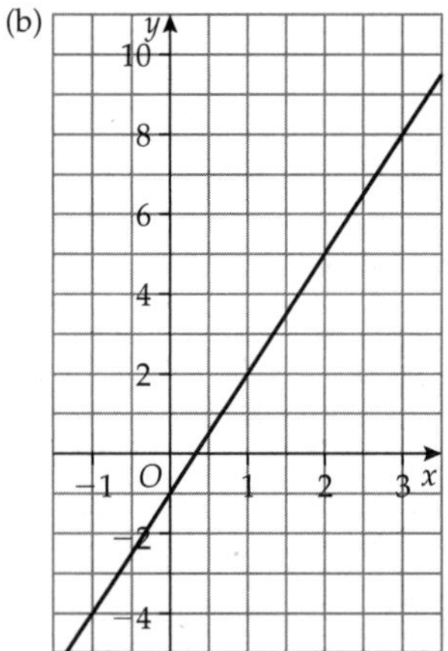

2.

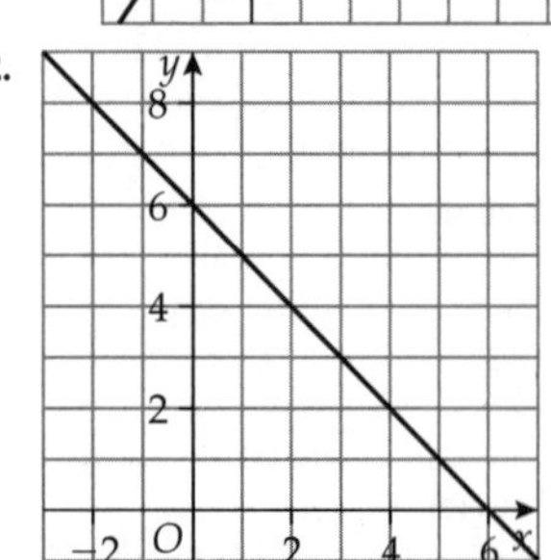

3. (a)

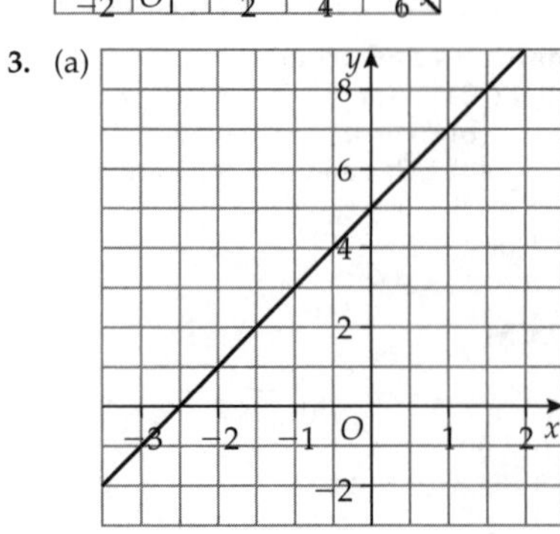

(b) 2

42. Real-life graphs

1. (a) 11 pounds
   (b) 6.8 kg
   (c) Liz: 10 pounds is 4.6 kg so 100 pounds is 46 kg
2. (a) 36 litres
   (b) 6 gallons
   (c) 20 litres is 4.4 gallons so 120 litres is 26.4 gallons

43. Distance–time graphs

1. (a) 10 km (b) 30 minutes (c) 11:40
2. (a) 30 km/h
   (b)

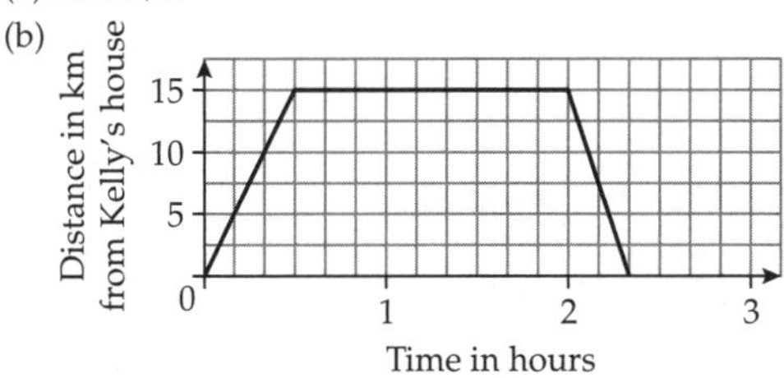

44. Interpreting graphs

1. (a) 2 pm
   (b) 7 am, 8 am, 9 am, 12 pm
   (c) 9–11 am
   (d) It decreased by about 2.25 °C
2. A and 3, B and 1, C and 4, D and 2

45. Quadratic graphs

1. (a)

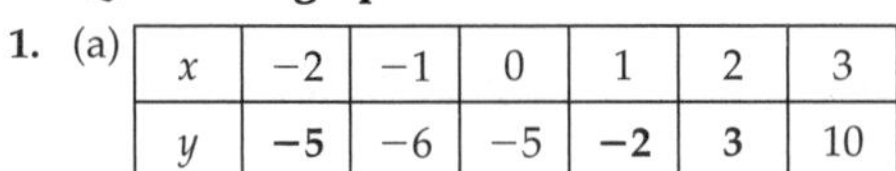

x	−2	−1	0	1	2	3
y	**−5**	−6	−5	**−2**	**3**	10

(b)

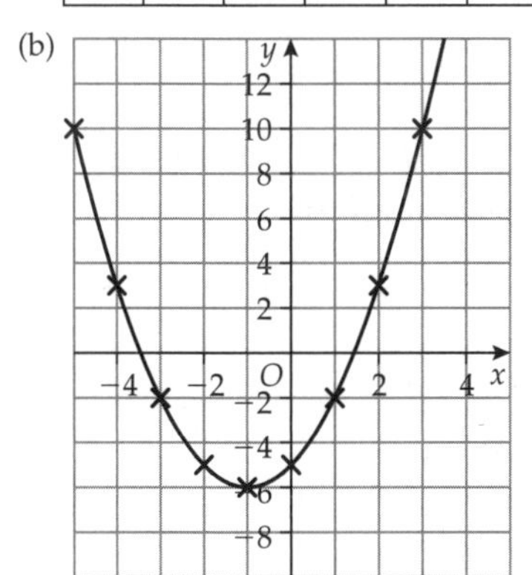

2. (a)

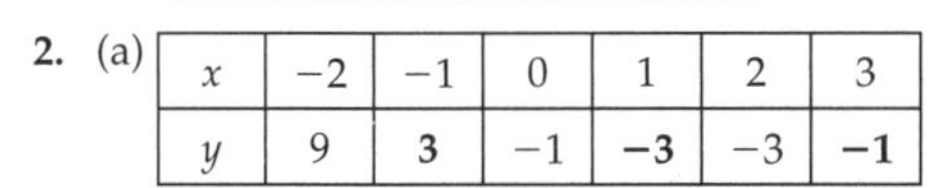

x	−2	−1	0	1	2	3
y	9	**3**	−1	**−3**	−3	**−1**

(b)

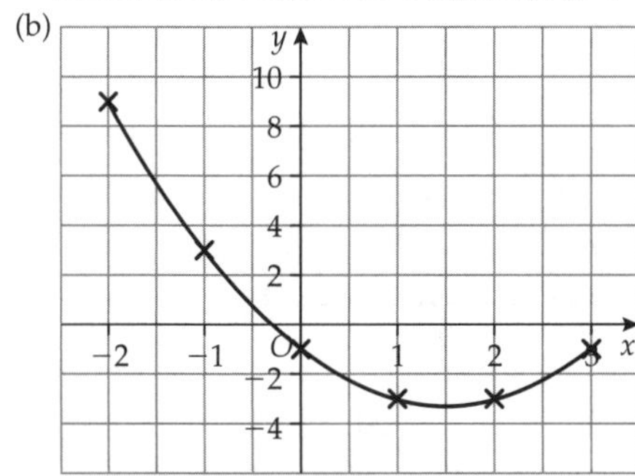

46. Using quadratic graphs

1. (a) −1.6, 3.6 (b) −0.4, 2.4 (c) −1, 3
2. (a) 3.2 (b) −1, 2 (c) −1.3, 2.3

47–48. Problem-solving practice

1. (a) 25
   (b) Yes, e.g. the sequence is 5, 9, 13, 17, 21, 25, 29, 33, 37, 41, 45, 49, … so the 12th pattern will need 49 sticks.
2. 70 pints
3. (a) £109 (b) 2.5 m
4. 89°

5.

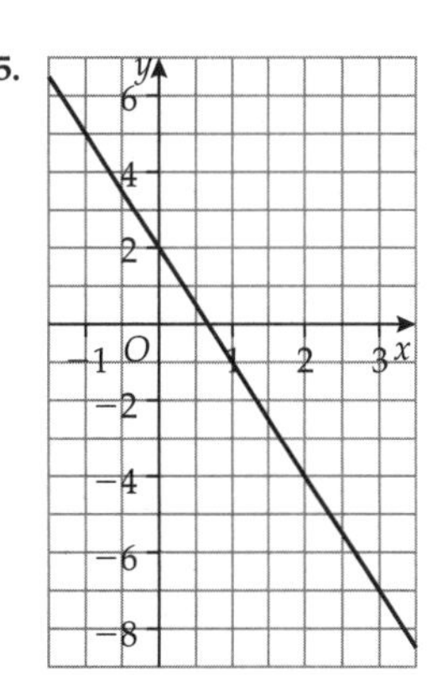

6. 5.5 cm

GEOMETRY AND MEASURES

49. Measuring and drawing angles

1. (a) 50° (b) 115°
2. (a) angle of 68° drawn (b) angle of 130° drawn

50. Angles 1

1. (a) obtuse (b) reflex
2. (a) 56°
   (b) angles on a straight add up to 180°
3. (a) 122°
   (b) angles around a point add up to 360°
4. The two angles should add up to 180 **not** 190; angles on a straight line add up to 180°.

51. Angles 2

1. 55° (angles in a quadrilateral add up to 360°)
2. (a) 127° (angles on a straight line add up to 180°)
   (b) 53° (alternate angles are equal)
3. 54° (alternate angles are equal)

52. Solving angle problems

1. 86° (base angles of an isosceles triangle are equal; angles in a triangle add up to 180°)
2. (a) 135° (corresponding angles are equal)
   (b) 50° (angles on a straight line add up to 180°; vertically opposite angles are equal; angles in a triangle add up to 180°)
3. 50° (angles on a straight line add up to 180°; angles in a triangle add up to 180°; base angles of an isosceles triangle are equal; angles in a quadrilateral add up to 360°)

53. Angles in polygons

1. 36° 2. 8 3. 24°
4. 18 5. 165° 6. 36°

54. Measuring lines

1. (a) 14 cm line (b) × at 7 cm
2. 7.8 cm or 78 mm
3. (a) 7.5 cm line (b) × at 3 cm from A
   (c) × at 6.4 cm from P
4. (a) e.g. 1.8 m (b) 5.4 m

55. Bearings

1. (a) 330°
   (b) Line drawn on a bearing of 200° from B
2. 320°
3.

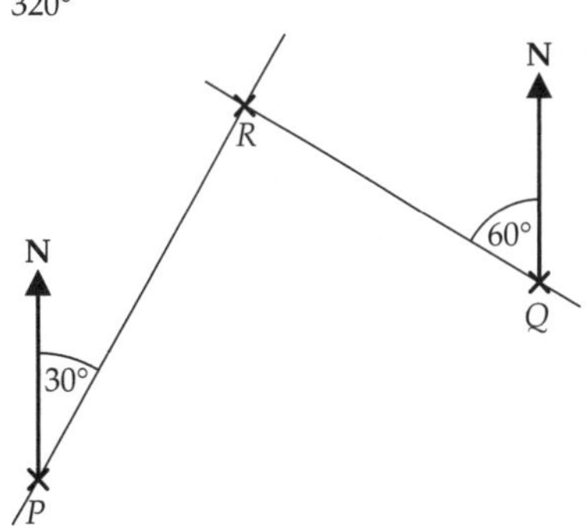

56. Scale drawings and maps

1. (a) 71 m (b) 11.9 cm
2. (a) 13.6 km (b) 35.9 km
   (c)

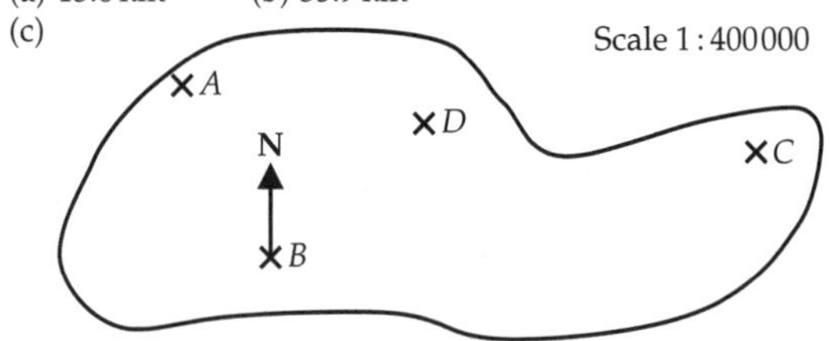

57. Symmetry

1. (a) 2 (b) 2 lines of symmetry drawn
2. (a) B, D (b) A, C
3. squares shaded

58. 2-D shapes

1.

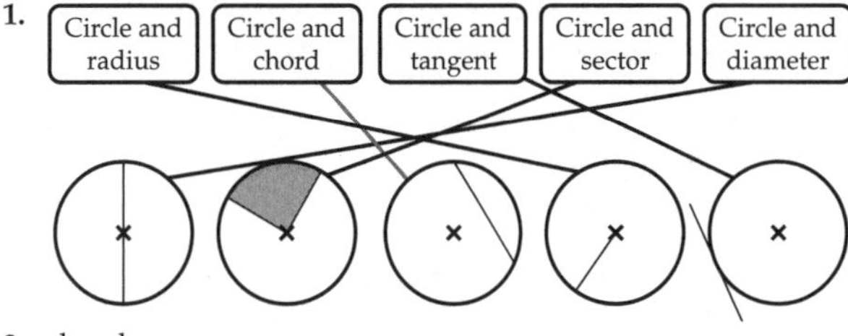

2. rhombus
3. (a) (i) trapezium (ii) kite
   (b) Parallelogram drawn
4. (a) A square has 4 equal sides and 4 equal angles.
   (b) Each angle in a rectangle is 90°.
   (c) A rectangle has 2 lines of symmetry.
5. square, kite, rhombus

59. Congruent shapes

1. (a) H (b) G (c) B
2. No, the shapes are different sizes
3.

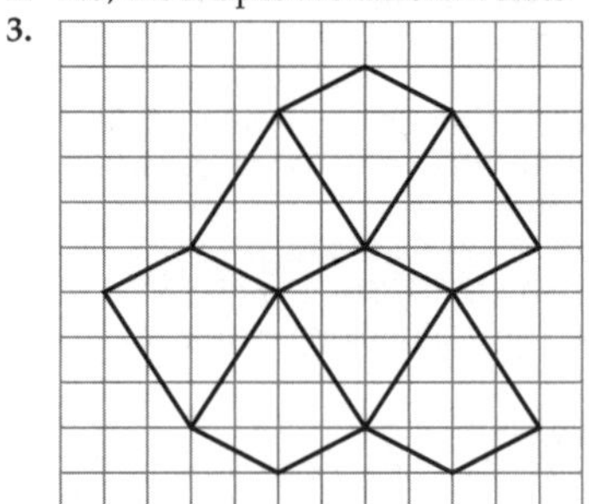

60. Reading scales

1. (a) 44 (b) 3.6
   (c) 420 marked (d) 3.8 marked
2. (a) 70 km/h (b) 160 m*l* (c) 32 °C (d) 4.6 kg

61. Time and timetables

1. (a) 18.10 (b) 2 hours 40 minutes
2. 07:20
3. (a) (i) 09:50 (ii) 2 hours 35 minutes
   (b) 07:08 (c) 10:30

62. Metric units

1. (a) 450 cm (b) 8.9 kg (c) 45 cm (d) 4080 m
2. 6
3. 0.4 g
4. 8 minutes 20 seconds
5. Cheaper at the supermarket

63. Measures

1. Mark 2. 40 miles 3. 56 km/h
4. £58.50 5. 2 hours 6. 5

64. Speed

1. 300 km/h 2. 200 km 3. 120 km/h
4. 1350 km 5. 4 km 6. 16:40

65. Perimeter and area

1. (a) 16 cm^2 (b) 18 cm
2. (a) 15 cm^2 (b) 24 cm
3. 31.6 cm 4. 25.4 m 5. 14 cm

66. Using area formulae

1. 40 cm^2 2. 15 000 cm^2 3. 20 cm^2
4. 2400 cm^2 5. 76 cm^2

67. Solving area problems

1. 72 cm^2
2. 6 tins
3. No; area is 5500 m^2 so he wants £22 000

68. Circles

1. 40.8 cm 2. 50.3 cm 3. 46.27 cm 4. 21.4 cm

69. Area of a circle

1. 153.9 cm^2 2. 266 cm^2 3. 226 m^2 4. 54.9 cm^2

70. 3-D shapes

1. (a) sphere (b) cylinder (c) square-based pyramid
2. (a) 6 (b) 12 (c) 8
3. cuboid drawn
4. (a) 8 (b) 18 (c) 12

71. Plan and elevation

1. cuboid with net 3; cylinder with net 4; prism with net 2; pyramid with net 1
2. (a)

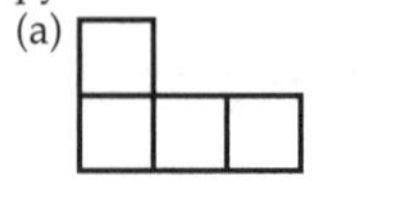

(b)

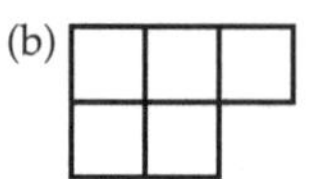

3. (a)

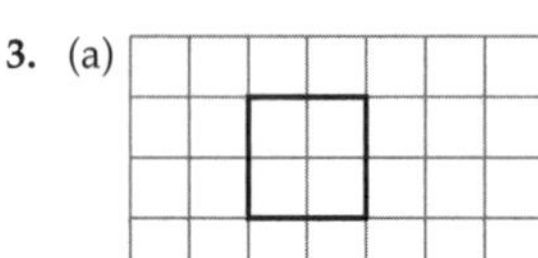

(b)

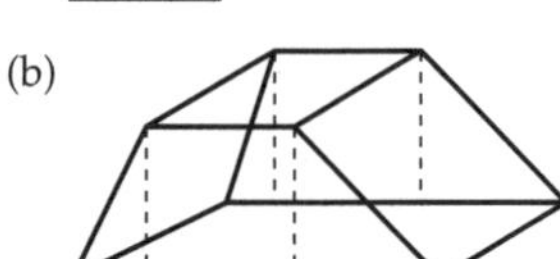

72. Volume

1. 21 cm^3 2. 120 m^3 3. 90

73. Prisms

1. 228 cm^2 2. 408 m^2 3. 412 cm^2
4. 360 cm^3 5. 320 cm^3

74. Cylinders

1. 1850 cm^3 2. 2040 cm^3 3. 2990 cm^2 4. 3 tins

75. Units of area and volume

1. 75 000 cm^2 2. 7000 mm^3 3. 4.2 m^3
4. 78.5 cm^2 5. 700 000 m^2 6. 9000 litres

76. Constructions 1

1. accurate construction
2. accurate construction
3. accurate construction

77. Constructions 2

1. accurate construction
2. accurate construction

78. Loci

1.

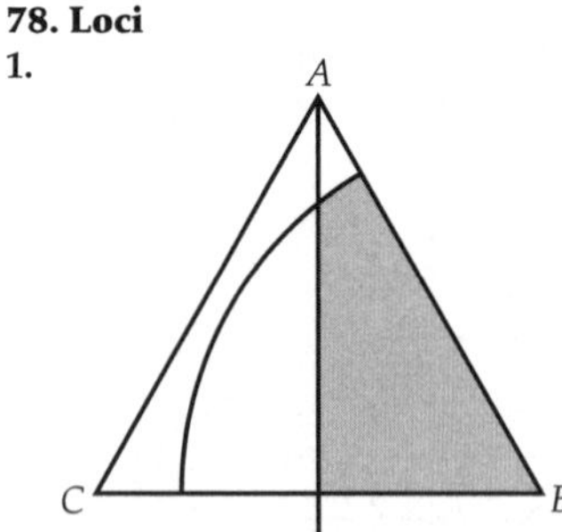

2.

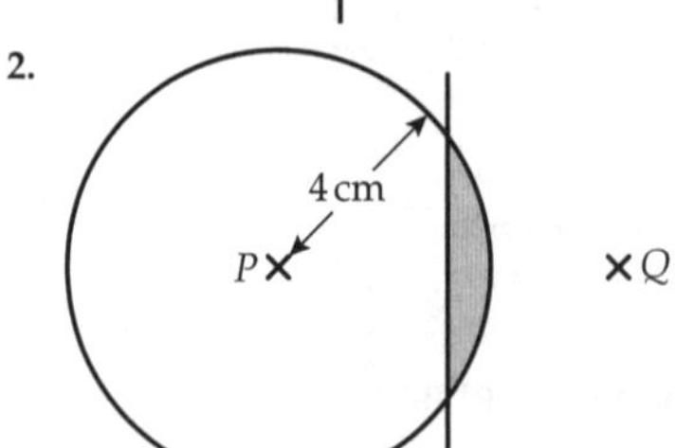

3. A B 2 cm 3.5 cm D C

79. Translations

1.

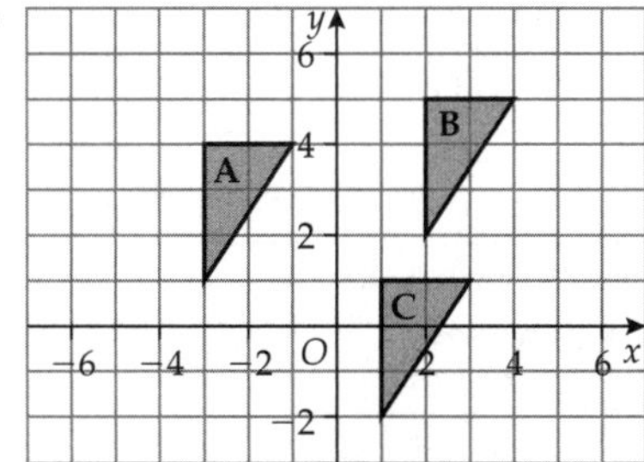

2.

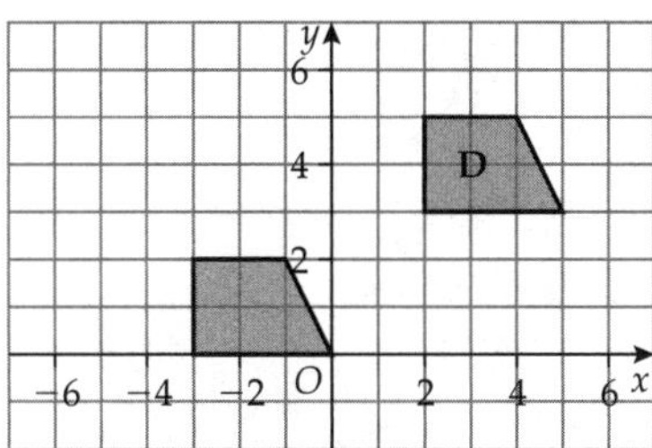

3. (a) translation of $\begin{pmatrix}-5\\2\end{pmatrix}$ (b) translation of $\begin{pmatrix}6\\-7\end{pmatrix}$

80. Reflections

1. (a)

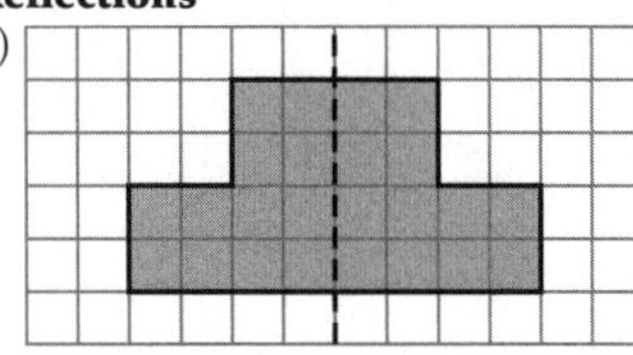

(b)

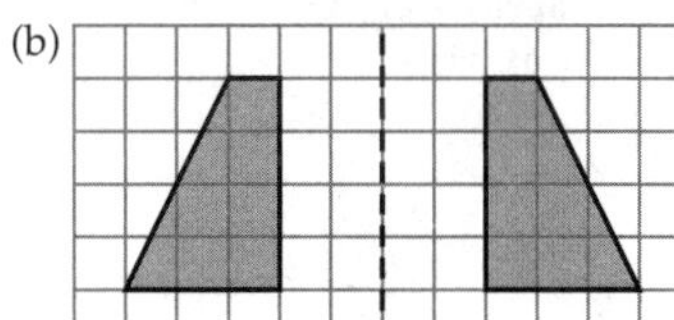

2.

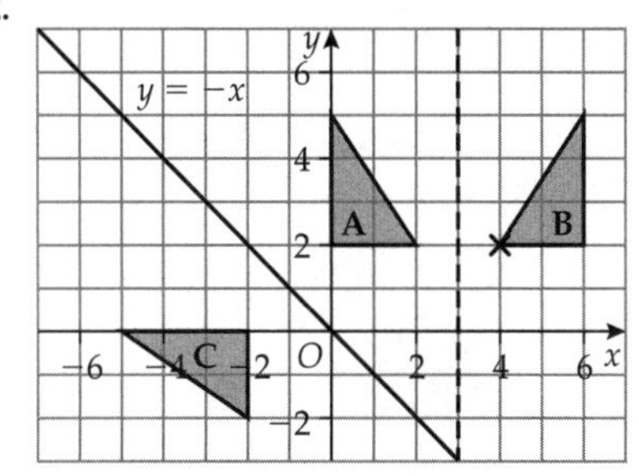

3. (a) reflection in the x-axis (b) reflection in $y = x$

81. Rotations

1.

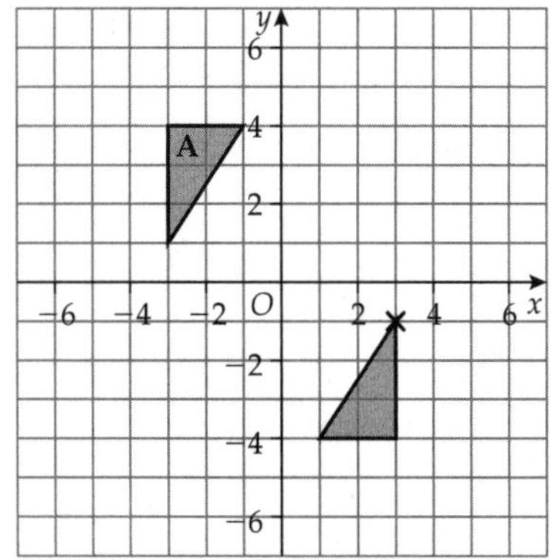

2.

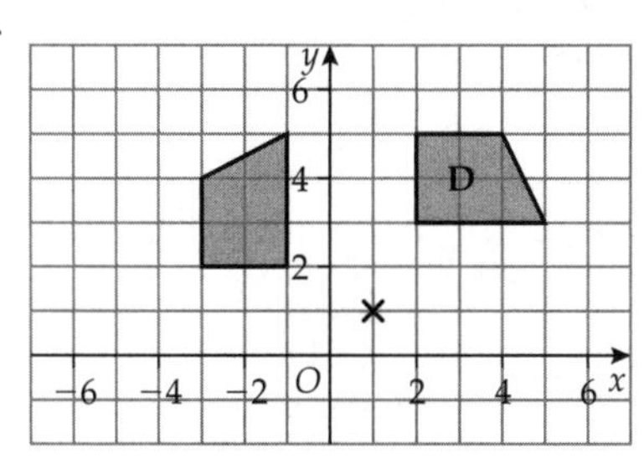

3. (a) rotation of 90° clockwise about (0, 0)
(b) rotation of 180° about (0, 1)

82. Enlargements

1.

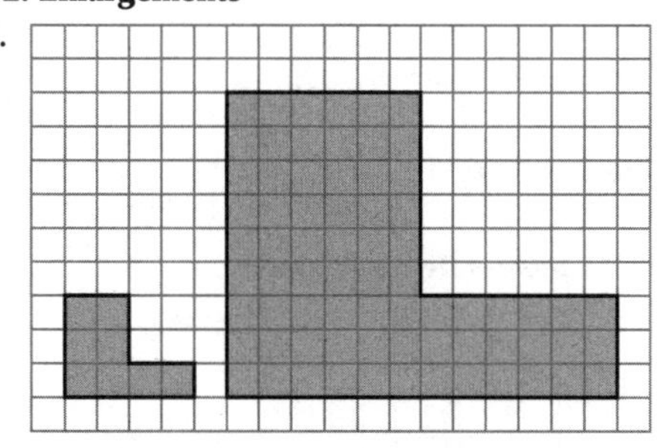

2.

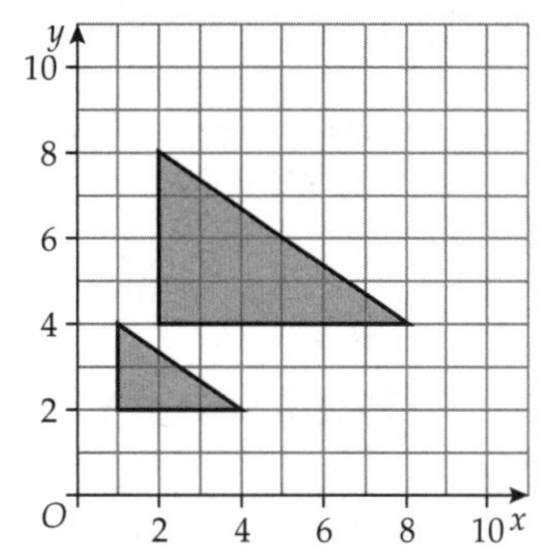

3. enlargement of scale factor 3, centre (1,0)

83. Combining transformations

1. (a), (b)

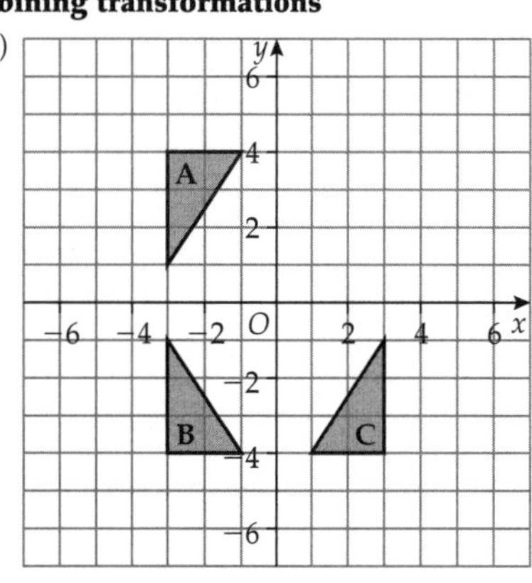

(c) rotation of 180° about (0, 0)

2. (a), (b)

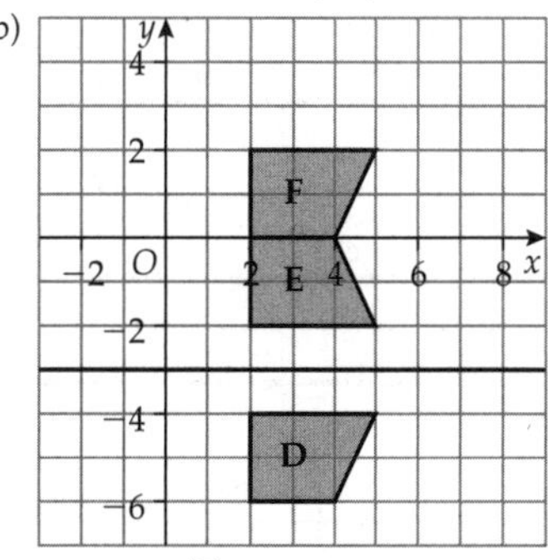

(c) translation of $\begin{pmatrix}0\\4\end{pmatrix}$

3. (a), (b)

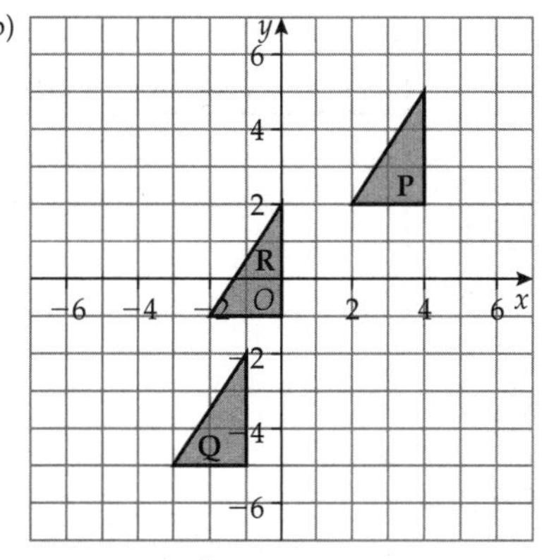

(c) Translation of $\begin{pmatrix}-4\\-3\end{pmatrix}$

84. Similar shapes
1. (a) D, F (b) B
2. 50°
3. Q

85. Pythagoras' theorem
1. 10.5 cm 2. 20.2 cm 3. 24 m 4. 22.2 km

86. Line segments
1. 13 units 2. 5.83 units 3. 15 units 4. 26 units

87–88. Problem-solving practice
1. e.g.

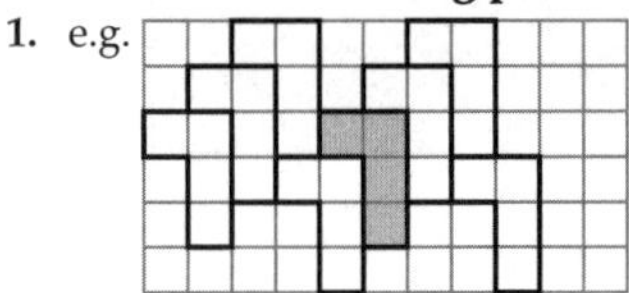

2. 624 cm^3
3. 69°; alternate angles are equal, angles in a triangle add up to 180°, base angles of an isosceles triangle are equal
4. 135 is not a factor of 360
5. £68.28
6. Rotation of 90° anticlockwise about (0, −2)

STATISTICS AND PROBABILITY

89. Collecting data
1.

Pet	Tally	Frequency
Cat		
Dog		

2.

Travel	Tally	Frequency
Car		
Bike		

3. Two from: no time frame, overlapping response boxes, no zero
4. How often do you go on holiday each year?
 0 ☑ 1 ☑ 2 ☑ 3 ☑ 4 or more ☑
5. How far do you travel to work?
 0–2 km ☑ 3–5 km ☑
 6–8 km ☑ more than 8 km ☑

90. Two-way tables
1.

	France	Germany	Spain	Total
Female	12	13	13	38
Male	21	12	9	42
Total	33	25	22	80

2.

	School lunch	Packed lunch	Other	Total
Female	16	6	17	39
Male	9	8	4	21
Total	25	14	21	60

3.

	Boys	Girls	Total
Liked coffee	13	11	24
Did not like coffee	2	14	16
Total	15	25	40

4.

	Jazz	Country	Other	Total
Male	30	8	9	47
Female	2	2	24	28
Total	32	10	33	75

91. Pictograms
1. (a) (i) 12 (ii) 7
 (b) 3 full circles drawn (c) $4\frac{1}{2}$ circles drawn
2. (a) 24 (b) 22
 (c) 4 large squares drawn (d) $1\frac{1}{4}$ large squares drawn

92. Bar charts
1. (a) 3 (b) Bus (c) 35
2. (a) 2.5 (b) Tuesday
3. (a)

	Tally	Frequency
Blue	𝍸 III	8
Red	𝍸 I	6
Yellow	𝍸	5
Purple	𝍸	5

 (b) fully labelled bar graph

93. Frequency polygons
1.

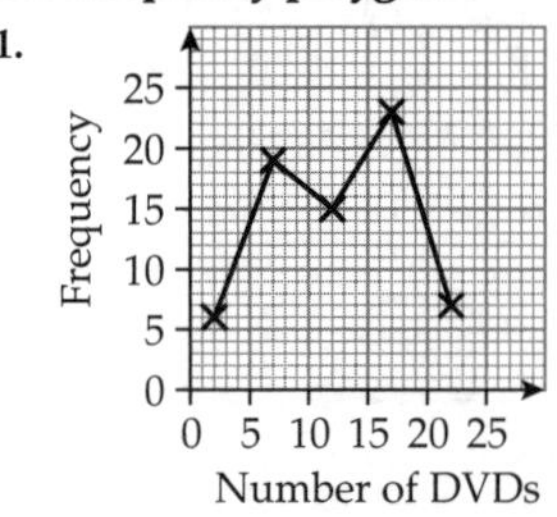

2.

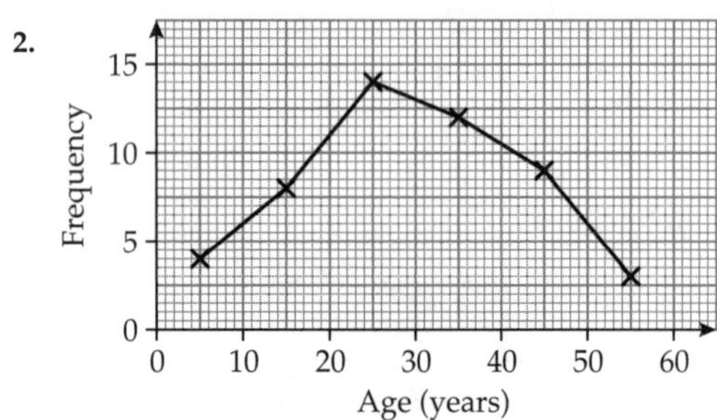

94. Pie charts
1.

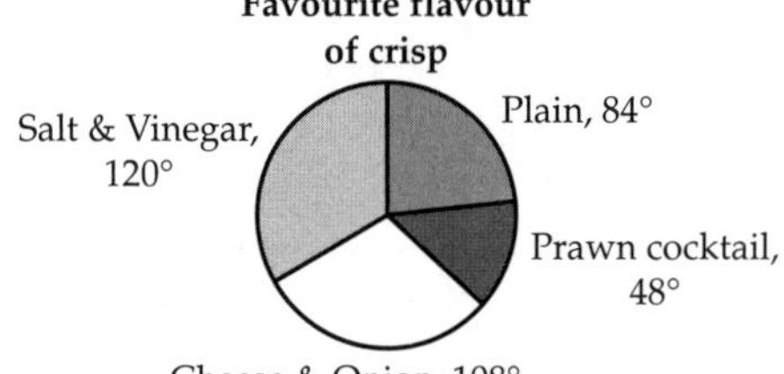

2. (a) Dog (b) 18 (c) 9
3. Drinks sold
 Hot chocolate, 72°
 Coffee, 144°
 Orange juice, 48°
 Tea, 96°

95. Averages and range
1. (a) 5 (b) 3.5 (c) 3.25 (d) 4
2. (a) 4 (b) 6.5 (c) 10.1
3. e.g. 5, 6, 10

96. Stem and leaf diagrams
1. (a) 31 (b) 24 minutes
2. (a) 25 (b) 18.5 (c) 27
3.

0	7 9
1	4 8 9 9
2	1 5 5 6 7 8 9
3	0 3 8
4	0 1 5 9

 Key: 4 | 0 represents 40 years

97. Averages from tables 1
1. (a) 0 (b) 1 (c) 4 (d) 1.15
2. (a) 17 (b) 16.5 (c) 16.4

98. Averages from tables 2
1. 15.25 minutes
2. (a) $70 < w \leqslant 75$ (b) $65 < w \leqslant 70$ (c) 69.1 kg
3. £45.50

99. Scatter graphs
1. (a) positive (b) line of best fit drawn
 (c) (i) 20 (ii) 32
2. (a) negative (b) £4000 (c) 3 years

100. Probability 1
1. (a) even (b) impossible (c) unlikely

2. (a) D at 1 (b) T at $\frac{1}{2}$ (c) E at zero
3. (a) $\frac{3}{16}$ (b) $\frac{5}{16}$ (c) 0
4. (a) $\frac{2}{7}$ (b) $\frac{5}{14}$

101. Probability 2

1. (C, T) (C, J) (C, W) (E, T) (E, J) (E, W) (H, T) (H, J) (H, W)
2. (1, T) (2, T) (3, T) (4, T) (5, T) (6, T) (1, H) (2, H) (3, H) (4, H) (5, H) (6, H)
3. 0.96
4. 0.2
5. (a) 0.49 (b) 0.31

102. Probability 3

1 18
2. (a) $\frac{1}{5}$ (b) 40
3. (a)

	London	New York	Total
Boys	26	12	38
Girls	15	7	22
Total	41	19	60

(b) $\frac{19}{60}$

4 108

103–104. Problem-solving practice

1. e.g.

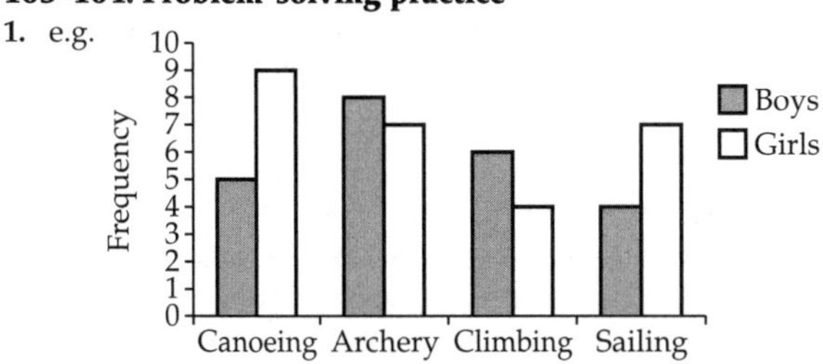

2. (T, H), (T, E), (T, P), (C, H), (C, E), (C, P), (W, H), (W, E), (W, P)
3. 7 people
4. 2, 2, 5, 6, 7
5. About 21–24 miles
6. 15 students

106–112. Paper 1 Practice exam paper

1. (a) 7 (b) Lion (c) 33
2. (a) 5026 (b) eighty (c) 3000
   (d) 0.09, 0.0971, 0.7, 0.79
3. (a) Rover (b) Volvo (c) Ford and Volvo
4. (a) Trapezium (b) F (c) 2 (d) 2
5. No, e.g. there is only room for 17 people.
6. (a)

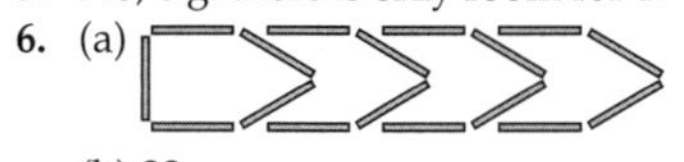

   (b) 33
   (c) No, e.g. all the patterns have an odd number of sticks and 80 is an even number.
7. e.g. 5, 11, 13
8. (a) Completed table (frequency column 2, 4, 4, 6, 2, 2)
   (b) 4
9. (a) (T, H), (T, E), (T, P), (C, H), (C, E), (C, P), (W, H), (W, E), (W, P)
   (b) £2.40
10. (a) gallons, kilometres, inches
    (b) 5000 g (c) 4.5 m
11. (a) Cardiff (b) 5 °C (c) 3 °C
12. (a)

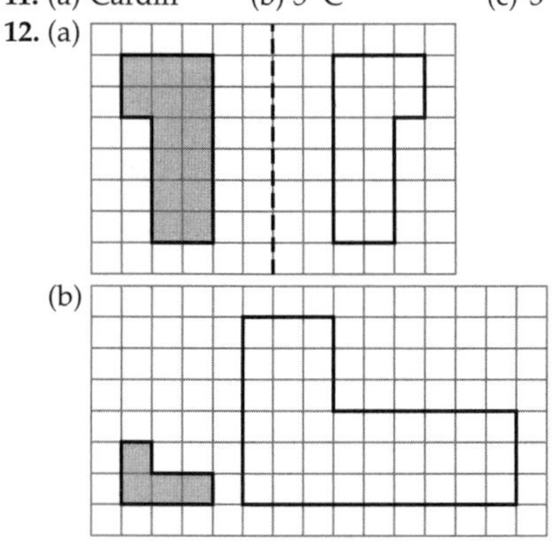

    (b)
13. 70°; angles in a triangle add up to 180° and base angles of an isosceles triangle are equal
14. (a) $\frac{6}{35}$ (b) $\frac{1}{3}$
15. 120 cm^3
16. $P = 32$
17. Accurate drawing
18. (a) $\frac{5}{12}$ (b) $\frac{2}{3}$ (c) 0
19. (a) $x = 7$ (b) $y = 36$ (c) $m = 4$ (d) $w = 3.5$
20. £24.60
21. 100
22.

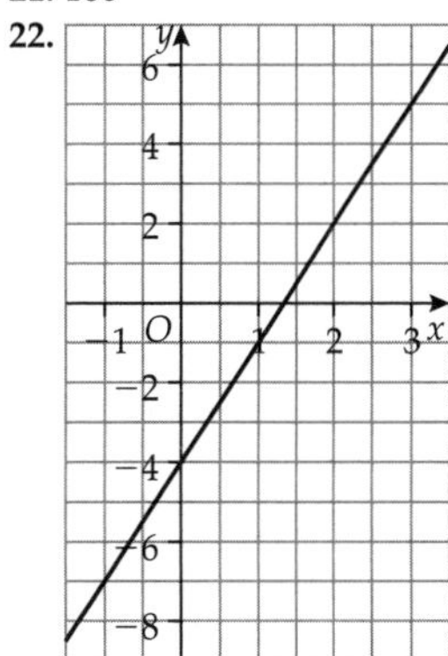

23. (a) e.g. The first two response boxes overlap; there is no provision for people spending less than £2.
    (b) e.g. How many magazines do you buy each week?
    □ 0 □ 1 □ 2 □ 3 □ 4 or more
24. £169
25. Accurate drawing
26. £43
27. 52

113–118. Paper 2 Practice exam paper

1. (a) 24 (b) 22
   (c) 2 complete rectangles for Thursday, $3\frac{3}{4}$ rectangles for Friday
2. (a) (i) (4, 3) (ii) (0, −3) (b) (−2, 3) plotted
   (c) (3, 1)
3. (a) $5c$ (b) $3x + 10y$
4. No, e.g. she is 25p short.
5. (a) 7.6 cm (b) (i) obtuse (ii) 135°
6. e.g.

Frequency; Boys; Girls; Swimming, Tennis, Athletics, Volleyball

7. 18 buses
8. (a) C and E (b) F (c) 2 lines of symmetry drawn
9. £56.88
10. (a) 5 (b) 3 (c) 3.3
11. (a) 0.75 (b) $\frac{1}{4}$ (c) $\frac{6}{7}$
12. 64.485 917 94
13. Shapes linked as follows (top row to bottom row):
    1 to 5, 2 to 4, 3 to 1, 4 to 3, 5 to 2
14. Pie chart with angles Jean 96°, Julie 120°, Harry 60°, Gwen 84°
15. £19.95
16. e.g.

Team	Tally	Frequency
Arsenal		
Newcastle United		

17. (a) 10 feet (b) 3.95 metres
    (c) 16 metres is approximately 4 × 13 = 52 feet so Kim throws the ball further.
18. (a) 665p (b) 26
19. 70
20. 300 g plain flour, 5 eggs, 500 m*l* milk, 125 g butter
21. £0.85 or 85p
22. 101 m^2
23. £530.40
24. Rotation of 90° clockwise about (0, 0)
25. No, he will need 11 tins.
26. $x = 5.6$

Published by Pearson Education Limited, a company incorporated in England and Wales, having its registered office at Edinburgh Gate, Harlow, Essex, CM20 2JE. Registered company number: 872828

www.pearsonschoolsandfecolleges.co.uk

Edited by Fiona McDonald and Laurice Suess
Typeset by Tech-Set Ltd, Gateshead

First published 2011

15 14 13 12
10 9 8 7 6 5

British Library Cataloguing in Publication Data
A catalogue record for this book is available from the British Library

ISBN 978 1 44690 014 7

Printed in Great Britain by Ashford Colour Press Ltd, Gosport, Hants.